Whale Songs and Wasp Maps

JOSEPH MORTENSON

Illustrations by Terry Magill

Whale Songs and Wasp Maps

THE MYSTERY OF ANIMAL THINKING

E. P. DUTTON NEW YORK

Published in the United States by E. P. Dutton,
a division of New American Library,
2 Park Avenue, New York, N.Y. 10016.

Library of Congress Cataloging-in-Publication Data

Mortenson, Joseph.
    Whale songs and wasp maps.

    Bibliography: p.
    Includes index.
    1. Animal intelligence. I. Title.
QL785.M67   1986   156'.3   86-4614
ISBN: 0-525-24442-5

Published simultaneously in Canada by Fitzhenry & Whiteside Limited, Toronto

COBE

DESIGNED BY EARL TIDWELL

10 9 8 7 6 5 4 3 2 1

First Edition

Grateful acknowledgment is given for permission to quote from the
following works:

All the Strange Hours: The Excavation of a Life, by Loren Eiseley.
Copyright © 1975 by Loren Eiseley. Reprinted by permission of
Charles Scribner's Sons.

For an adventurer I know
named Leif

CONTENTS

ACKNOWLEDGMENTS

This book was written in many libraries. Its first draft was started in Nova Scotia in the libraries of Dalhousie University and Mount St. Vincent University. Its basic concepts were refined in Massachusetts at the public libraries of Ipswich and Boston. The manuscript was finished at three libraries in Sonoma County, California—those at Sonoma State University, Santa Rosa Junior College, and the town of Sebastopol. I would like to thank the many librarians who helped me on this long, westward journey. I would also like to acknowledge the writers of all the books on animals I read in the collections of these fine libraries.

Whale Songs and Wasp Maps might have remained in my indecipherable longhand were it not for the support and encouragement of several people. For their generous help, I owe thanks to Robert and Joyce Abel, Neil Ehrlich, Hazel Flett, Joyce Gronroos, Nancy K. Innis, and Edward L. Walker.

Whale Songs and Wasp Maps

Plus ultra ("More beyond")
—MOTTO ON THE
PILLARS OF HERCULES

1

ON SONOMA MOUNTAIN

The rains broke early on the coastal range this year. Coming slowly at first, and then faster, warm fall storms hurled themselves at the high ridges and draws, filling the dry creeks and renewing the springs. Just as the nights grew longest, the procession of storms ceased. Yesterday the temperature fell sharply, and dense winter fog began to rise inland and on the shore. But above this cold obscuring dampness, the coastal ridges rode like vast ships upon a flat, placid sea of mist. Tonight, the night sky over the ridges has cleared for the first time in weeks, and Halley's comet threads dimly through the sky.

High on Sonoma Mountain, a flock of sheep stirs as a wan yellowish light appears in the east. Rising one by one in the unaccustomed cold, the sheep stand and stretch. Dark wet patches in a heavy hoarfrost mark their resting places. The light in the sky slowly brightens and an old white-muzzled ewe starts to move down the rounded slope. All the others follow, save one small gray sheep. For the first time in a year she turns away from the flock.

Restless, the gray ewe moves hesitantly up the field. Stopping by a fenceline, she sits awkwardly, with her spine curved and her head raised toward the dawning sky. Water spills from deep within her body onto the hoarfrost. The ewe's breath steams and the water steams. For half an hour she twists and strains. Then, after a heave, two small hooves point from her rump. The ewe pushes again and a small head appears, pressed against knobby forelegs. Soon a wet white lamb tumbles outward onto the crystals of the frost.

The gray ewe turns and begins to lick the newborn lamb. The lamb raises its head, blinking in the light of its first day. Within minutes the lamb stands, its body still dripping, and fumbles at the gray ewe's udder, tasting her new, custard-thick milk.

It is a good day to be born.

In the woods at the edge of the field, I sleepily approach the ewe. Startled by my footfalls in the duff, a varied thrush flies and alights on a live-oak branch. True masters of the woods are these thrushes that search every yard of its ground and catch intruders like me at our every false step. We eye one another for a moment before I move on. I see the sheep but keep my distance. The ewe is cleaning the standing lamb, and the lamb is sucking. In these first hours of life, a bond is being set between ewe and lamb. Once the ewe learns to know her lamb, she will reject all others. The ewe will keep the lamb near her and search for it if it strays. All is well with them; there is no need to intervene.

Anyone who keeps sheep on these hills walks in their world. We brought the merino, the Suffolk, the Dorset, the Corriedale, and all the other breeds to these coastal ranches, but these sheep are the true owners of the ridges and canyons. They know the grasses for each season and the shelters from the storms. We can follow their footpaths to water or to the safety of the heights. These paths can lead us to ravines and knolls that we have never known. On their familiar grounds the sheep travel close to one another in flocks, watching us fearfully on our rare intrusions into their domain.

What are these beings that we shepherds nourish, guard, and sacrifice? What is created in the cycle of breeding and birth that we oversee?

I have lived among sheep long enough to know that they are like me. In summer, they seek the noon shade of the live oaks and race to taste cold water. In winter, they walk on the rounded slopes out into the warming sunlight. Yet between the sheep and me is a barrier as immense as the space between us and the sun.

Though the sheep are my familiar companions in these hills, I cannot enter their inner world. I cannot feel the fear of a sheep, even though I may be its cause. I have never experienced the bond that moves the ewe to bleat after her lamb, though I, too, may be searching for the stray.

In my isolation, I can even doubt that the sheep are conscious. They might be mere mindless bodies on a mindless mountain, fit only to be constrained and consumed. But in my heart I know this is false—the sheep are as conscious as you or I.

More than this, I believe that the sheep and I are not the only ones aware of these ridges far above the mist. The varied thrush that returned my gaze must be conscious, too, as must its cousins the robins, pacing in the short-cropped winter grass. I believe this, although I have not shared the sentience of a thrush watching in the woods or of a robin hunting in the field.

THE LIMITS OF EXPERIENCE

Which of us has known his brother? Which of us has looked into his father's heart? Which of us has not remained forever prisonpent? Which of us is not forever a stranger and alone?
                                          —THOMAS WOLFE, Look Homeward Angel

What keeps us apart? What keeps me out of your mind and the mind of the sheep? I cannot share the consciousness of other beings because there is an impenetrable wall between all of us. This wall is the world of matter, a world that is ruled by the laws of physics. This wall is the wall of nonexperience.

The only matter we convert into experience is the matter within our nervous systems. All the rest is matter that we can never experience directly. The great exterior world can act on us only by causing the cells of our nervous system to fire. Our contact with the outer world is limited to the impulses it excites in our nerve cells.

What we receive from the world outside our experience is a steady flow of messages. The external world excites our sensory cells, which then pass electrochemical impulses up our nerves. Each of these pulses is identical. When I look at Sonoma Mountain, the sunlight reflected by the mountain falls on special receptor cells at the back of my eyes. These sensory receptors initiate a series of pulses that travel along pathways toward my brain. At this level there is no Sonoma Mountain; there are only messages in a simple code. Two signals are possible: a pulse or silence. Yet it is out of this crude binary information that the brain constructs its own version of the world outside itself.

We ourselves are neither conscious of the individual pulses in the incoming messages nor of the impinging energy that initiated the pulses. We become conscious of incoming sensory information only after it is analyzed at higher levels. It is in my brain that I recreate an image of the fields and oaks on the rounded slopes of Sonoma Mountain. This image is what I experience directly. I do not experience Sonoma Mountain itself, but rather the re-creation of it within my brain.

THE SEARCH FOR ANALOGY

All psychic interpretation of animal behavior must be on the analogy of human experience.

—MARGARET FLOY WASHBURN, Animal Mind

Why do I believe in your consciousness, reader? You are even more remote from me than Sonoma Mountain. But you are familiar to me. Your body, your brain, your actions are like my own. Since your appearance in the world of matter is so much like mine, I assume that you have a consciousness that is like my own. Of course, our brains and our behavior differ in all kinds of detail and so our consciousnesses will not be identical. But we are physically similar enough that I am convinced that your consciousness is in some essential way like my own. Why else would I write these words for you?

If you and I are conscious, what of the sheep? And what of the thrushes that visit us in winter? Perhaps they are conscious. Perhaps they are not. How can we tell? The answer is not immediately given us.

As with ourselves, the nervous systems of these animals are hidden beneath the masks of their bodies. The brains and nerves of vertebrates lie beneath skin, muscle, and bone. We can see these bodies move—but we cannot see whether a consciousness underlies the movement. Yet even if we cannot directly see a consciousness, there is a way to seek whatever may be hidden behind the façade of an animal's body.

To search for consciousness in others, we must first turn to ourselves. We are the one animal in which we can indisputably relate consciousness to the body. We are learning more and more about the brain processes and structures that underlie our own awareness. Suppose that we look at other animals in the light of this new information about the bases of our consciousness. Are the critical structures and processes that underlie our awareness possessed by other animals? If the sheep on Sonoma Mountain have a brain that is like ours in general structure, then we might feel that there is a clear possibility that the sheep are conscious, since we know our brains are the source of our own consciousness. If we look further and find that identical processes within the sheep's brain or our own control whether sheep or people appear alert, then the argument that sheep are conscious becomes more compelling. The greater the likeness between the sheep and us, the more convincing the inference of consciousness will be.

What happens when we find differences in brain and behavior between ourselves and some other animal? Consider the ked, a small whitish insect often found beneath the wool of sheep. If we studied the nervous system of this wingless fly, we would find that it was quite unlike our own; thus we might feel it was less probable that the ked had a consciousness like our own, or that it had a consciousness at all. If we watched keds crawling through the wool of sheep, we would discover that their behavior was quite different from ours, as well: we don't walk on six legs or breathe through the sides of an external skeleton. The greater the number and kinds of differences we found, the weaker would be our inference that the ked's mind existed or was like our own.

My own interest in animals began far away from Sonoma Mountain. I spent two years in a darkened laboratory watching South American knifefish. My knifefish rested quietly by day but swam gracefully by night. The fish emitted a steady stream of weak electrical pulses, which I amplified and played through a speaker. Some of the fish sounded like tuning forks and some like buzzers. The fish could use these signals to detect each other or objects in their tanks. Hearing the signals gave me access to their sensory world and I was entranced. Yet I did not think beyond the signals to the intelligence that they served.

Only when I left the laboratory did I wonder about a mind that lay beyond behavior. On Sable Island, ninety miles off the shore of Nova Scotia, I found I had no wish to experiment on the harbor seals that we were to observe. There was consciousness in their eyes, in their care for their pups, and in their vigilance against the figure with the telescope on top of the dunes. I didn't want to intervene in what I didn't fully understand. Yet if the seals were conscious, how could I study such an impalpable thing?

This book represents my answer. In it we will search for similarities and differences between ourselves and other life-forms. Any similarity we find to those processes essential to our consciousness will suggest that the other life-form may be conscious. Any difference that we find will imply that the other life-form is not conscious or has a different kind of consciousness. This method is straightforward: we are simply looking for essential analogies between ourselves and other beings. This is our means of looking for whatever lies beneath the feathers of the thrush and the wool of the sheep.

It is ironic that such great effort has been made to scan the galaxy for evidence of intelligent life. Arrays of radio telescopes track distant stars, listening for any sign of extraterrestrial life. We broadcast specially designed messages into space so that someone may hear us. No extraterrestrial intelligence answers. Yet our earth abounds in life, and the signs of terrestrial intelligence are there for those who seek them.

To search for consciousness is truly a grand adventure; we are looking beyond the world of matter to seek whatever sentience may be there. On our way we must be careful not to become lost in the theories of physics, or to be stopped by the material surface of things. We must remember that we are looking through skin and skeleton to an inner reality. At the same time we must not carelessly venture too far from our observations of brain and behavior into easy and baseless speculation. It is these data about the material world that will form the common ground for our search. Whatever we infer must be clearly tied to this hard material base; if we disagree, we can see plainly where our interpretations vary.

In our search we must be humble, more like pilgrims than conquistadors. We must not forget that there is a limit to our knowledge of other consciousnesses. This limit is ourselves. If we cannot experience a state, we can never know it. If a thrush's feelings are like our own, then we can understand what

those feelings might be like. But if a thrush's feelings are utterly
unlike what we experience, we will never know them.

We can only feel what we can feel, sense what we can
sense, think what we can think.

On Sonoma Mountain, the gray ewe and her new lamb are
moving down to join the flock. As the days grow longer, other
lambs will be born on the high fields. The mountain will re-
sound with the reedy bleat of the lambs and the deeper voices
of the answering ewes.

Sonoma Mountain lies pleasantly now, its colors growing
warmer in the light of late afternoon. The mountain was born
in volcanic fire, four million years ago. It was conceived in the
collision of continental plates and formed out of upwelling mol-
ten rock. How could I ever feel the heat of its birth or sense the
enormity of its mass? All I can do is to create my image of the
mountain, a pale human abstraction of the qualities of its existence. Yet the pleasure and colors of Sonoma Mountain are
mine; the mountain has neither heart or eyes. Or should I say
that it is merely the limits of my kind of knowledge that I
perceive when I look at Sonoma Mountain? What is Sonoma
Mountain? Is it myself? Is there more beyond?

*The Swallowes which at the
approach of springtime we see to
pry, to search and ferret all
corners of our houses; is it
without judgment they seeke, or
without discretion they chuse
from out of a thousand places,
that which is fittest for them, to
build their nests and lodging?*
　　　—MONTAIGNE, Essays

*And, doubtless, when swallows
come in spring, they act in that
like clocks.*
—DESCARTES,
　*from a letter to the
　Marquis of Newcastle*

2

THE LOST TRADITION

What is reality?

What is a stone? A leaf? An octopus in its den?

What is a swallow on the wing?

We may think that we know what the stone is, yet a physicist will tell us the stone is really a thin vapor of electrons, protons, neutrons, and other, more fleeting bits of mass. If we press the physicist about this invisible mist of subatomic particles, he begins to tell us about a mysterious miniature world of enormous power and force. The physicist's stone becomes more like the philosopher's, creating the extraordinary out of the familiar.

We may think we know what a barn swallow is: a tireless airborne insect-catcher with a cinnamon buff breast and dark blue back. In spring, we may first see them swooping over the

ponds. Later we see these fork-tailed fliers wheeling in and out
of the barn, as they carry mud to make nests on the rafters out
of our reach. But is this bold swallow only the graceful body that
we admire in flight? Like the stone, the swallow may be far
more than we can see on the surface.

I believe that the consciousness of each animal is like a
reality within a reality, a world within a world. The consciousness of bird and fish is like our own consciousness, a universe
that encompasses the universe within ourselves and outside
ourselves. Yet, like the mist of particles that constitutes the
physicist's stone, the consciousness of all animals is invisible.

The invisibility of consciousness forms a great barrier to
understanding. It isolates us all, one from another. One reaction
to this barrier is to deny consciousness. This was the tack taken
early in this century by John Watson in psychology and Jacques
Loeb in biology. These men believed that science should study
reflexlike behaviors and shun the language of consciousness.
Their behavioristic views became the dogma of academic psychology and biology in the 1920s. Watson wrote:

> Behaviorism claims that consciousness is neither a definite
> nor a usable concept. The behaviorist, who was trained always as
> an experimentalist, holds, further, that belief in the existence of
> consciousness goes back to ancient days of superstition and magic.
> The great mass of people today has not yet progressed very far
> away from this savagery—it wants to believe in magic.

In the halls of universities, animals came to be regarded as
things without thought, as objects without reflection. For two
generations of students, the white rat and frog were held up as
models for our self-understanding. Science had lost its mind.

I can remember fellow students in my college psychology
classes laughing at somone who was foolish enough to ask a
question about his mind. The rest of us had certainly become
sophisticated enough never to think about our own.

It is strange that a concept as fundamental as mind can go
out of fashion in much of academic science. Mind is existence.
Mind is experience. Mind is at the center of what we are. Ulti-

mately, all science is an explanation of mind. The concept of mind can only be neglected and scorned when scientists believe that they can explain, predict, and control the things that are important to us solely in physical terms. Unfortunately, the fact that one can be precise about the details of some simple physical system can give us the illusion that we can predict and control events within ourselves. Somehow it is easier to cling to limited certainties about matter than to pay attention to the wider world of consciousness. Yet in doing so, the scientist can become irrelevant to us.

Twenty years ago the purely "objective" world of the protected academic scientists began to be breached. Jane Goodall went to the Gombe Stream Reserve in Tanzania to see how chimpanzees actually lived in nature. Dian Fossey climbed the Virunga volcanoes in Rwanda to watch and defend the last mountain gorillas. Sloops sailed out of St. John's and Honolulu, dragging hydrophones to hear whales in the depths of the sea.

The great apes began to force open the door that the behaviorists had shut on animal consciousness. Their rich social and emotional life became intelligible to us, once we had been patient enough to see how they actually behaved in the jungle and rain forest. People also began to think about intelligences in the sea, intelligences that were both like and unlike our own.

Scientists began to rediscover animal mind. The first surveys of animal consciousness in decades were written. Donald Griffin, the pioneer who, as an undergraduate, had first described the ultrasounds of bats, published *The Question of Animal Awareness* in 1975. Griffin revised this book in 1981 and has published an even more recent survey in 1984, entitled *Animal Thinking.* Griffin also organized an extensive collection of articles entitled *Animal Mind—Human Mind* in 1982. The English psychologist Stephen Walker wrote a readable review called *Animal Thought,* published in 1983, that covers some of the diverse scientific literature. A new vessel was being launched in science. Scientists were setting sail for an unknown shore that was as invisible and yet as real as thought.

In this way, a revolution has begun in psychology and biology. Scientists are starting to go beyond the visible, beyond the body, and beyond behavior to seek whatever consciousness may be there. But how can we go beyond the objective? How can we study the invisible? How can we know what is hidden within the bodies of other animals? The physicist looks beyond the surface of the stone; how can we look beneath skin and bone to find whatever reality lies within?

Before the rise of behaviorism, there was a venerable tradition of scientific concern with animal mind. Charles Darwin believed that evolution formed both bodies and minds, and wrote an entire book on the expression of mind. Darwin's student George Romanes was the first to work out a system for inferring minds. C. Lloyd Morgan and others refined this approach and made careful observations of creatures as diverse as amoebas and dogs. Yet this tradition was lost in the triumph of behaviorism.

It is time now to take up the Darwinian tradition again. Behaviorism let us impoverish our world and make animals into mere objects. The world is so much richer than this, full of animals that are subjects. To deny the consciousness of other animals is to deny the true wealth of the world, to turn away from the very reality of ourselves and creatures like ourselves.

Margaret Floy Washburn was one the last adherents of this lost Darwinian school of thought. Washburn was a professor at Vassar College and wrote a text called *Animal Mind* that went through three editions, the last of which was published in 1926. Washburn was well aware of the strengths and deficiencies of the methods used to study animal mind. She recognized the inaccuracy of dramatic and unsubstantiated anecdotes as a source of knowledge about animals. But she also saw that the limited experimental conditions of the laboratory led to misleading simplifications of animal behavior. Washburn favored long and objective observations of individuals, the kind that were later made of the apes.

Washburn devoted a chapter of her book to the use of analogy for inferring mind. She noted that we could find suggestive parallels in either behavior or brain structure between ourselves and other animals. If a certain behavior accompanied our consciousness, then other animals showing similar behaviors might be conscious. If an animal was similar to us in brain structure, this too might imply consciousness. Finally, Washburn mentioned that physiology was a third possible source of analogies for inferring consciousness.

THREE ANALOGIES
FOR INFERRING ANIMAL MIND

BASIC OBSERVATION	TYPE OF ANALOGY	SOURCE OF INFORMATION
Brain	Structure	Anatomy
Nerve process	Function	Physiology
Behavior	Consequence	Psychology/ethology

In her book, Washburn included a historical review of the ways in which animals were perceived and pointed out that there have always been two extreme views. For reasons of religion, philosophy, science, and, one suspects, personality, some people have treated animals as if they were mindless machines, without real or significant sentience. At the other extreme are those people who have interpreted animals in the most human terms possible. Washburn took the middle course between these poles. She was conservative in her interpretations, yet liberal in her acknowledgment of the varied and perhaps unknowable possibilities of animal consciousness. I adopted her outlook, which represents the culmination of the traditional study of animal mind. She would have been pleased, I think, to realize how far we can now go with her logical approach and the latest scientific evidence.

Cleave a piece of wood, I am there; lift up the stone and you will find me there.

—SAINT THOMAS, THE APOSTLE, *Saying 77*

3

THE THREE KINGDOMS

Look around. Turn your senses outward. Energy from ten thousand sources is arriving at the specialized sensory cells of your body, which are the gates of your experience. Out of this flood of incoming information you fashion images of the physical world: mice, men, pebbles, mountains, blades of grass. What a triumph of existence, your sheer awareness of all these raw sensations and your organization of this information into images, with which you order the external world.

But where does awareness itself begin in this universe that we contemplate from the narrow windows of our senses? Is awareness a prize that we alone possess? Or is awareness a more primitive quality, more democratically distributed than we recognize? Perhaps we should first consider the least obvious source of awareness: the "inert" matter that forms the backdrop for the lives of plants and animals. By ancient convention, this is the mineral kingdom: the sun, the stars, the air, the water, and the rock of earth.

THE MINERAL KINGDOM

[I]n the world there is nothing to explain the world . . . nothing to explain the necessity of life, nothing to explain the hunger of elements to become life, nothing to explain why the stolid realm of rock and soil and mineral should diversify itself into beauty, terror, and uncertainty.

—LOREN EISELEY, All the Strange Hours

Most of us do not believe that there is consciousness in minerals. Our ancestors may have worshiped the sun and stars as capricious gods, and they were awed by the powers of benevolent waterfalls and springs, but we have lost any such faith. We no longer speak with the mountains. Who could conceive of a consciousness residing in a stone or a stream? Who could communicate with a lake or a thundercloud?

None of these lifeless physical things resembles us. In them there is nothing like our bodies, our brains, our behavior. Since they bear no likeness to our physical selves, we cannot believe that consciousness exists in them. We are, after all, the known physical standard of consciousness. We know we are conscious and we know what we look like; that much is certain. But therein lies a dilemma. We are limited precisely by what we look like and by what we experience. While we can reason that something that looks like us might feel like us, we cannot reason the opposite way. When we find physical things that are unlike ourselves, we simply have no way of drawing conclusions about their consciousness. We have no way of knowing. Yet there is no logical reason to limit consciousness to our own life-form or to other life-forms like us.

In the seventeeth century the philosopher Leibniz proposed that the world is made up of monads. In essence, each monad is a mind; its primary attribute is that of perception. Monads have many curious properties, but it is their consciousness that is most interesting. The consciousness of monads varies in its clarity; different monads have different degrees of sentience or awareness. Monads are like mental atoms out of which matter is constructed; according to Leibniz, all matter is thus perceptual or mental. How quaint this view sounds today. We know that atoms don't have sentience. Mere matter can't feel or perceive. Or can it? We really have no evidence either to deny or affirm that an atom or a molecule might have some form of consciousness. How would we know?

There is, of course, a single exception to our general rule of ignorance about consciousness and matter. The exception is ourselves. We know that consciousness must spring from the matter that constitutes us. Quarks, leptons, protons, neutrons, atoms, and molecules are the very stuff of our brains. At some level—subatomic, atomic, or molecular—experience must arise. And if experience can arise out of the matter in us, where else might it surface?

As we shall see in what follows, one physical process that is clearly involved in consciousness is electricity. Electricity in our bodies depends on ions, or charged particles. Ions are created when an atom or molecule gains or loses an electron or two. Ions that have lost electrons are positively charged; ions that have gained electrons are negatively charged. If we put table salt, or sodium chloride, into a warm glass of water, the salt dissolves into two ions: the positive sodium ion and the negative chloride ion. Because of the mixture of the ions in the water, there is no buildup of a net electrical charge. A glass of salt water won't shock us. However, if there were some way of separating the positive and negative ions, a voltage would develop. The voltage would reflect the shortage of electrons in the positively charged sodium ions and the excess in the negatively charged chloride ions. Electrons, or current, would flow between the unequal charges if it were possible. In our bodies, the walls of our cells often serve to separate ions and to generate voltages. Consciousness involves such voltages, and simple ions such as those of sodium and chloride form an atomic ground for awareness.

But at what level does consciousness emerge? Are we conscious of the actual ions at the level of atoms and molecules? Or are we only conscious of more complex kinds of electrical voltages or currents? We are certainly conscious of something material—but what? If we only knew when or how consciousness came out of matter, our search for mind in the universe would not be so blind.

THE VEGETABLE KINGDOM

A child said What is the grass? *fetching it to me with full hands; How could I answer the child? I do not know what it is any more than he.*

—WALT WHITMAN, Leaves of Grass

Plants are not totally unlike us. They are made of cells, and these cells contain protoplasm and genetic material, just as our own cells do. The cells of plants are organized into specialized groups and structures, just as are the cells of our own bodies. Plants belong to species, which are groups that interbreed and reproduce in nature. These species can be classified into families, orders and classes, and these groups all have an evolutionary history. We humans, too, constitute a species and have an ancient family history. These parallels between the plants and ourselves link us together as living beings.

Plants differ from ourselves, however. Most plants are nourished by minerals that are obtained directly from the soil and by the carbohydrates that they create from water and carbon dioxide through photosynthesis. The majority of plants do not move freely, but are fixed to one site. Most animals do move about, and are all ultimately the consumers of plants. Unlike plants, we animals cannot create our own food out of earth, light, air, and water. But there are exceptions among plants. Some plants are parasitic. A few others increase mineral uptake by capturing animals. Pitcher plants, found in nutrient-impoverished bogs, lure insects into a slippery basin of digestive fluids. Yet even these exceptional plants are not mobile like most animals.

Although most plants do not move from their sites, they are not motionless. Plants exhibit several kinds of behaviors. Many of these are not obvious to us, since plants typically behave in slow motion. The first widely recognized behaviors of plants were *tropisms*, which were first described in the nineteenth century. A positive tropism is a movement that is directed *toward* a stimulus, and a negative tropism is one that is directed *away* from it. For example, many plants orient their leaves and flowers toward the sun in daylight. The turning of the face of a sunflower toward the sun is a common example of positive phototropism or orientation to light. The shoot of an oat seedling will move away from the direction of gravitational pull even when it is grown in darkness, which is a movement called negative geotropism. More complex than these simple orientations are the contact tropisms seen in climbing vines. The tendrils of the scarlet runner bean wave in slow circles until some object is encountered; then the tendril wraps itself around the object.

More dramatic to us are those few plant movements rapid enough to be seen. The leaves of the sensitive plant, *Mimosa pudica*, collapse when touched, which is presumably a defense against browsers. Among the insectivorous plants, the well-known spring of the Venus flytrap, *Dionaea muscipula*, is rivaled by the suction release of the bladderworts, or *Utricularia* (Slack 1981). The bladders of these odd rootless plants develop a partial vacuum: insects or larvae, drawn to nectar lures, trip special trigger hairs and are sucked up to be digested within the bladders.

In addition to movements, plants show many physiological responses to stimuli. For example, many plants do not accept being eaten by animals without protest; they poison herbivores with a nasty assortment of chemicals that can impair digestion or reproduction, destroy RNA or DNA, or drug the nervous system. Several plants increase such poison production when their leaves are attacked (Schultz 1983). In the eastern United States, gypsy moth larvae feed on the young leaves of red oaks. Oaks and other trees can respond to such leaf injuries by increasing the production of tannins and other poisons in the uninjured leaves. Since the carbohydrate production of the leaves is partially diverted into poison-making, this defensive reaction is not without metabolic cost. Thus it makes evolutionary sense if increased poison production is reserved for actual insect attacks.

Poison-making is not obvious to us, since it is both slow and internal. Yet even more subtle is the communication that can pass between nearby plants about insect attacks. Recently it has been learned that uninjured willows will produce poisons when neighbors are attacked. This communication appears to depend on an airborne chemical given off by the besieged neighbors.

Any plant that responds must also be sensitive. Even though we rarely think of the sensitivity of plants, in each field and forest there must be millions of sensors responding to sunlight, gravity, water, and other stimuli that direct plant growth. How do plants show such sensitivity if they lack obvious eyes or organs of balance? And how can these sensors guide the response of the plant, if the plant has no nervous system?

Charles Darwin and his son Francis were the first to isolate sensors in plants. The Darwins found that it was the extreme tip of the leaf sheath in canary grass and oat seedlings that was sensitive to light. Conversely, they determined that it was the very tip of the root that was positively sensitive to gravity. In the 1920s, Frits Went of Utrecht discovered that the bending of a shoot to seek sunlight is the result of the flow of *auxin*, a substance that regulates plant growth. Auxin will flow differentially within the leaf sheath of a grass seedling when the source of light is shifted to one side. The auxin moves to the dark side of the grass stem, accelerating the growth of that side and thus bending the whole stem to the light. Auxin was the first of a

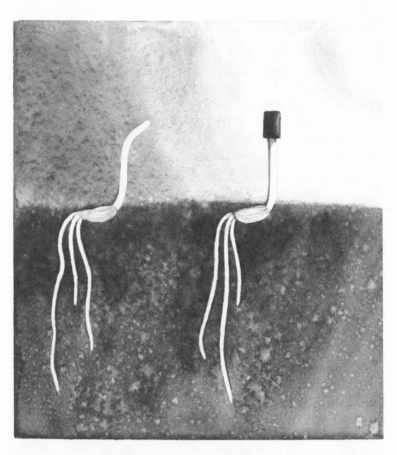

Oat seedlings seeking light. When the Darwins covered the tip of a
seedling with a darkened glass tube, it no longer bent toward the light.

number of plant hormones to be discovered. These regulatory
chemicals control many plant growth patterns and movements.
Small quantities of these controlling chemicals have powerful
effects, which is the case in animals as well.

Hormones may have more than one function in a plant.
Auxin, for instance, also helps direct the positively geotropic
root tips. If a potted plant is turned on its side, its roots begin
to bend at right angles downward toward the new direction of
gravity. In such a plant, auxin concentrates in the bottom side

of the now-horizontal root tip. In contrast to its effect in the tip of the shoot, auxin appears to inhibit the growth of the cells at the bottom side of the root. Since the cells on the top side thus grow relatively faster, the root turns downward.

What are the receptors in the shoot and root tip that respond to light and gravity? In the case of the shoot tip, the light receptor seems to be a carotene photopigment. Photopigments are substances that change in response to light conditions, and carotenes are red pigments such as those found in carrots. Sunlight can cause this particular photopigment to change its form, and this change, in some as-yet-unknown way, releases the differential movement of auxins. In the root tip the suspected receptors are the *statoliths*. Statoliths are free-falling grains of starch or other particles that can move within their cells. The theory is that the position of the statolith against the cell membrane determines the direction of the root growth.

Even greater sensitivities have been claimed for plants. A book called *The Secret Life of Plants* (Tompkins and Bird 1973) cited several instances of apparent telepathy in plants. The first case was that of the dragon tree, *Draceana massangeana*. Lie detector expert Cleve Backster attached a galvanometer to the leaves of a dragon tree that decorated his office. He used the galvanometer to measure electrical resistances. If someone is threatened or frightened, skin resistance will drop, and a galvanometer can detect such a change. In theory, resistance also drops because of guilt, and a galvanometer can thus catch a person in a lie, but in fact the drop could be caused by another strong feeling, such as anger at the interviewer. Backster decided he would try to produce a resistance change by threatening his dragon tree. Unexpectedly, the resistance of the dragon tree changed at the moment Backster *thought* of burning its leaves. The dragon tree seemed to read his mind.

Unfortunately, there are two difficulties with Backster's tale of telepathy. The first and most severe one is that most investigators have failed to make similar observations: the electrical resistances and potentials that can be recorded in plants simply do not appear to change in response to human thoughts.

Unless some way is found to demonstrate telepathy in repeated experiments, it can't be studied or interpreted.

The second difficulty is in the interpretation of the electrical potential. If you can detect a typical biological voltage in the leaf of a philodendron, what does this mean? Does one assume that this *biopotential* is a sign of consciousness? Or is the biopotential merely a routine sign of life? Or might it be that the voltage isn't really a sign of anything other than an ionic imbalance, such as can be observed in batteries or power outlets?

The interpretation of biopotentials is not a simple matter. Certainly such voltages accompany life in animals, and the absence of biopotentials signifies death. In some cases we can explain what specific patterns mean, as in the case of the potentials generated by the heart. In other cases, the meaning of the biopotentials is not completely clear, as in the case of brain waves. Yet, even though the meaning of brain biopotentials may not fully be understood, these voltages certainly accompany our own consciousness. What can we say about the biopotentials of plants?

Potentials in plants originate in the membrane that surrounds each cell. This membrane permits certain substances to pass into the cell and restricts the movements of others. Water and carbon dioxide can readily permeate the membrane, while many ions will be resisted. The membrane may also actively pump some minerals and sugar into or out of the cell. Other substances can either enter or leave the cell, depending upon the state of the membrane.

In and around plant cells are ions such as those of sodium and chloride. If a cell membrane can act as both a pump and a barrier, then such ions can be selectively transported across a membrane and become concentrated within the cell. If more negatively charged ions come to be concentrated in the cell, the inside of the cell will be electrically negative to the outside. This is exactly what happens in nature; many kinds of cells in plants and animals have a negatively charged cell interior. Since cells are small, the negative charge produced is also small—on the order of millivolts. Yet, considering the size of most cells, these voltages are quite respectable.

What observations have been made about these biopotentials in plants? One well-studied case is that of the freshwater alga called *Chara* (Scott 1962). The cells of this green alga were selected for study because they are quite exceptionally large— a single cell can be 1.5 millimeters in diameter and 15 centimeters long—roughly the dimensions of a pencil lead. Their size means that it is not too difficult to specify ion concentrations and to relate these concentrations to electropotentials. Within the cells of *Chara* there are relatively high concentrations of positive potassium, sodium, and calcium ions. The positive charges of these ions are offset, however, by an even greater internal concentration of negative chloride ions.

Perhaps the most curious thing about the cells of *Chara* is the electrical pulse that is seen when the cells are poked or are stimulated chemically. Such a stimulus causes the internal negative potential to drop briefly to one-quarter of its normal value. This brief pulse then travels slowly down the length of the cell. The normal resting potential is restored following its passage. During the passage of the pulse there is a change in the cell membrane's permeability: calcium ions flow inward to reduce the potential, and then potassium ions flow out to restore the original level.

The meaning of the traveling pulse in *Chara* is not established. As we will see later, however, this pulse bears an uncanny resemblance to signals in the nerves of animals. This suggests that the behavior of plants or algae could be coordinated by brief voltage pulses. The electrical response seen in *Chara* might also help explain some early stage in the evolution of nervous tissues in animals. Such primitive electrochemical processes would surely be involved in any possible consciousness developed in plants. Interestingly, rapid pulses have also been recorded from carnivorous plants such as the Venus flytrap and sundew (Benthrup 1979).

In many other plants, slower electrical responses can be

recorded from groups of cells, such as those found in stems or root tips. Relatively long-lasting voltages and current flows have been found during plant growth and movement. For example, when a grass stem bends toward light in a positive phototropism, an electrical potential develops across the shoot (Scott 1962). This potential change accompanies the movement of the growth hormone auxin in an electric wave. Since the speed of the wave is only about a half inch an hour, it is not fast enough to qualify as a pulse. On the other hand, it illustrates the relatively slow time scale of many plant responses.

How are such long-lasting electropotentials related to ion flows such as those seen in *Chara*? Let us consider a kind of ionic bedtime story to illustrate the kinds of relationships that have been found recently. This is the story of the "sleeping" movements of pea family plants like the silk tree, *Albizzia julibrissin*, and the rain tree, *Samanea saman*. These plants have leaves that are held horizontal or open during the daylight hours and that assume a vertical or closed position at night, which is the opposite of our own behavior, of course. The ionic and electrical events in this "sleep" have been studied intensively (Racusen and Satter 1975; Satter 1979; Galston 1981). The opening and closing of the leaves are the result of swelling in special motor cells in the pulvinus, an organ at the base of the leaflets. Potassium and chloride ions flow into the upper motor cells in daylight; water then diffuses into the motor cells, enlarging them and making the leaves lie flat or open.

At night the process is reversed: potassium and chloride flow out of the upper side of the pulvinus and into the lower side. The lower side then swells with water and the leaves close for the night. The flow of the potassium and chloride ions is in part regulated by the photopigment phytochrome. It is suspected that this particular photopigment can cause rapid changes in the permeability of cell membranes. This appears to be the case in the sleeping movements of the leaves of the rain tree at nightfall. The membrane potential of the lower pulvinar cells drops with leaf opening in daylight and increases with leaf closure in darkness. This biopotential shift precedes the changes in ion concentrations, supporting the view that phyto-chrome causes membrane electrochemical processes that lead to leaf movement and ion transport.

All of the bioelectric phenomena in tropisms and leaf movements are reliable; anyone can repeat these observations and obtain these data. From this evidence it is clear that elec-tropotentials in plants are linked with growth and movement. While the exact chain of causation underlying all movement and growth is not yet determined, ion flows and biopotentials may well provide many of the links between sensors, hormones, and responses in plants.

In summary, it can be said that plants move without mus-cles, see without eyes, and control themselves without a ner-vous system. Plants accomplish these feats without any exotic means; chemicals are transformed, ions flow across membranes, and hormones run through cells. These processes are familiar to us since they lie at the base of our own growth and activities —yet they appear strange, because plants seem so unlike us.

If there is sentience in plants, I suspect it is tied to their coursing ions and hormones and to the associated generation of electric potentials and currents. This is what is common to plants and to our nervous systems. But this is an all-too-human view, a search for our very image among the plants; there is no reason why their consciousness or the physical processes under-lying it should parallel our own. Why should either the sun-flower or the sun be made in our image? I have no answer; I am uncertain that I have asked the right questions of the plants.

We have considered the hard material data from which consciousness in plants might be inferred. If plants lack sen-tience but are instead mindless automata, then these data are just the record of a biological machine. They stand only for chemicals and voltages—nothing more. If, however, plants are sentient, then these same chemical transformations, ion move-ments, and biopotentials may be the very source of conscious-ness.

THE ANIMAL KINGDOM

When we watch an animal how are we to know whether it is con-
sciously thinking or merely computing; whether it is a sentient being,
or simply an unconscious well-programmed robot . . . ?
    —GOULD AND GOULD, *"The Insect Mind: Physics or Metaphysics"*

Vast is the kingdom of the animals: there are over two million
kinds. Their origin seems to lie in marine plankton. The first
multicellular animals to leave traces in sedimentary rock were
the odd, flat Ediacarians, more than 650 million years ago (S. J.
Gould 1984). The animals diversified in the sea and entered
fresh water; they ascended from water onto land and into the
air. Finally, in the last 5 million years or so, members of our own
genus appeared on the savannas of Africa. At some point during
this evolutionary flow, perhaps even at its outset, mind appeared and then evolved into the form we now experience.

How can we seek the origin of mind amidst this great
diversity of organisms, among the clams and trilobites, the
pterodactyls and whales? What was the evolution of our consciousness? There is a way to order all the beasts and thus to
seek evidence for mind. Let us look at the kind of cells that
underlie our minds and acts, and then trace their appearance
and combination in the kingdom of the animals. These are the
cells of our nervous systems and our muscles.

In our nervous systems we find billions of nerve cells or
neurons. Typical animal neurons always remind me of spindly
trees. From the cell body extends a long trunklike *axon.* The
axon has a few terminal branches. From the cell body also
extend shorter *dendrites,* which branch out like finely divided
roots. While neurons may look something like trees, they act
like rechargeable batteries. The cell interior has many more
potassium ions than the outside, which has a greater concentration of sodium and chloride ions. These differences in concentration are created by the continual expulsion of sodium ions
and intake of potassium ions. The cell membrane also traps

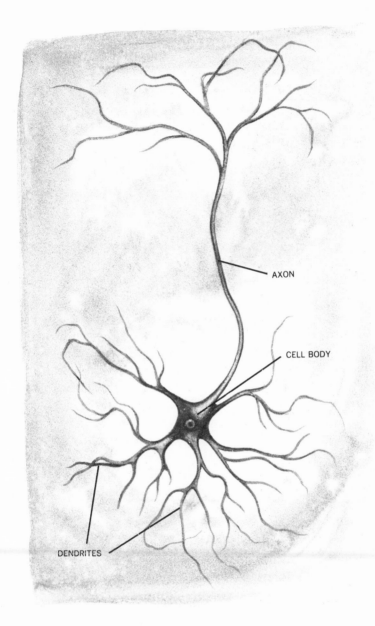

AXON

CELL BODY

DENDRITES

Neuron. Impulses are typically transmitted from an origin in the dendrites or cell body outwardly along the axon and its branches.

large negative protein ions within the neurons. These large ions make the overall inside charge negative.

When a neuron is stimulated sufficiently, it "fires" or discharges at the point of excitation. There is a change in membrane permeability, and sodium ions rush into the neuron. This produces a momentary reversal of polarity, the inside of the neuron briefly developing a positive charge. This impulse or "spike" of electrochemical activity then travels down the neuron. As the spike passes, the neuron begins the process of recharging itself with an outflow of potassium ions. Thus, most of the time the neuron acts like a recharging battery; when it is fired, however, the neuron acts like a wire conducting brief electrical pulses.

Neurons never exist in isolation; they are linked with other neurons and sometimes with sensory receptors, muscles, glands, or other effectors. The link between any of these cells is called a *synapse*. The synapse is actually a very narrow gap between a transmitting neuron or receptor and a target neuron or effector. At most gaps, the transmitting cell releases small packets of chemicals that enter the target cell. There are several kinds of transmitter chemicals; some excite the target cell and others inhibit the target. If there is adequate excitement or depolarization, the target cell will fire. The transmitter causes the membrane permeability to change so that the negative cell voltage begins to fall near the synapse. If the voltage falls far enough, an impulse is initiated that then travels along the target neuron, muscle, or other element. In the case of inhibition or hyperpolarization, the target cell is made less likely to fire. The transmitter substance alters the membrane so that the negative voltage of the target cell is actually increased. Thus depolarization is more difficult and firing less likely.

Neurons are the basic units of our nervous systems. Neurons communicate with each other through synapses, and this is how much of the business of our brains is conducted. In a sense, neurons are the basic physical units of mind, the biological elements within our bodies that are the basis for experience.

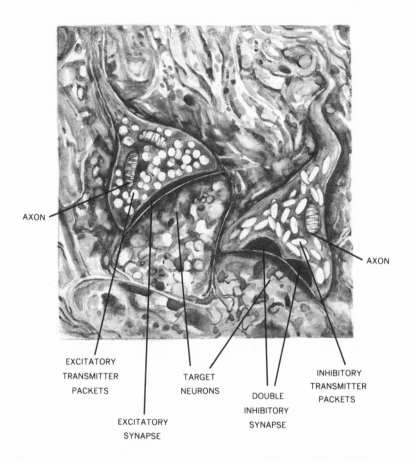

AXON

AXON

EXCITATORY
TRANSMITTER
PACKETS

TARGET
NEURONS

INHIBITORY
TRANSMITTER
PACKETS

DOUBLE
INHIBITORY
SYNAPSE

EXCITATORY
SYNAPSE

Synapse. At the synapse, arriving impulses release small packets of neurotransmitter. The release of transmitters into this gap between the neurons makes firing in the receiving cell either more or less likely.

Although we consciously do not seem to sense individual nervous impulses, we can detect activity in a single neuron or small group of neurons. For example, our eyes are so sensitive that they can detect about seven quanta, or units of light energy. This minimal amount of light should stimulate only an equal or smaller number of photoreceptor cells, which should in turn stimulate only a few incoming neurons. While we seem only to become conscious of this activity at higher levels of the brain, we still are dealing only with a small proportion of cells out of the billions in the brain.

We cannot see the neurons in our body, except in a single case. If we look at the back of the eyeball—that is, if we peer into someone's eye—we can see the retina in the back of the eye. The top layer of the retina is composed of sensory neurons. These neurons are sending visual information from the back of the eye to the brain. Even though we can see these transmission lines, we cannot see anything happening, since the impulses traveling through the nerves are electrochemical.

Although we cannot see electrical events in neurons, we can see the effects of neural events on our bodies. Impulses are sent along nerves from the brain and spinal cord to reach our muscles. Walking, talking, typing, and many other behaviors are the result of direct brain commands. Of course, the action we see is not the set of commanding impulses itself, but the copy of these impulses made by the muscles. In this way, behavior makes our nerve impulses visible.

Behavior is the means by which the brain can attempt to control the world around it. The brain's control can also extend within our bodies. The brain's most immediate internal control is through muscles and glands; our nervous system can act on the inner world either by making muscles contract or by causing secretions to flow.

Our power over the inner and outer world thus rests on contractions and secretions. Some of our power is conscious. We can consciously will ourselves to act, and so alter the world. Yet obviously the powers of consciousness are limited; we cannot command the sun to stop or wish away the mountains. We cannot even control all kinds of basic events within our own bodies. What exactly can consciousness control? What is beyond our conscious will? What is under conscious control?

This question is a critical one. If we do something unconsciously, then it is possible that other animals might also. If we can only do something consciously, then it is likely that other animals may be conscious if they match our actions.

Consider our glands, so important in the control of our internal environment. Normally we do not go around directly willing our glands to secrete. We do not will our sex hormones to flow, nor do we command our stomach to secrete its digestive enzymes. Instead, these flow of their own accord. In other words, the immediate control of these secretions is unconscious. Thus, if we found that other animals secreted sex hormones or stomach enzymes, we would not consider that as evidence that they were conscious. On the contrary, we would suspect that these actions were unconscious. One should add, however, that we *can* influence changes in many types of secretions, even though our control over them is not conscious. If we make love or eat, our glands can obviously be affected indirectly by our acts. Other animals can do the same.

What about muscles? What are our conscious powers over our muscles? What muscle action is under unconscious control? In humans, muscles are of two basic types: *smooth* and *striate.* Smooth muscles are comprised of a number of separate spindle-shaped cells, each with its own nucleus. Smooth muscles are relatively short and their action slow and sustained. The power of a smooth-muscle contraction depends on the number of individual cells contracting. We usually do not observe such contractions, since smooth muscles are not attached to the skeleton and thus do not cause conspicuous movements of the body. In other words, the consequences of smooth-muscle actions are internal and not external. Control of most smooth-muscle movements is typically unconscious or involuntary. We do not normally will the muscles of our blood vessels to contract or relax, nor do we control the movement of food within our gut.

Striate muscles get their name from their distinctive dark and light bands, or striations. Striate muscles are relatively long and are made up of a number of fiber bundles. Each bundle consists of a number of separate fibers. These individual fibers originated when a mother cell repeatedly divided into a number of daughter cells. The outer membranes of these cells disappeared, leaving a single long fiber with many nuclei. When we make a muscle, a number of these muscle fibers contract. The harder we make the muscle, the more fibers contract. The response of striate muscles is typically fast and precise. Since these muscles are ordinarily anchored at one or both ends to the skeleton, their effect is often external. Movements of skeletal muscles are usually easy to observe: we see people stand or run, sit or stretch, and all of these represent the movement of striate muscle and bone.

Our control of skeletal muscles is direct; that is, we can consciously choose to stand or run. Traditionally, striate muscles are called *voluntary*, since their movement can be subject to our will. Our control over striate muscles is typically expressed at the level of coordinated acts. For example, we usually do not command individual muscles of our leg to move; most of us don't even know their names. Instead, we choose to walk, run, or skate, and are not at all conscious of the commands being sent to individual muscles. In fact, we often tend to direct our conscious thinking at the outer environment. We think, "I will skate across the frozen river," and do not worry about the individual muscles, but rather about the ice below us. Of course, when originally we learn to skate or walk, we may be paying a lot of attention to the muscles involved, as well as to the ice or the sidewalk.

There are exceptions to the rules about the conscious control of skeletal muscles. For example, if a goshawk suddenly loomed out of the woods near us, we might duck before we had a chance to think. This primitive reaction to a suddenly looming object is reflexive in its control, and does not involve a conscious decision to avoid a collision. Another class of exceptions is exemplified by those bodily movements seen in sleep or during other states we would not call conscious.

Exceptional kinds of control are also possible over smooth muscles. For example, it would be a messy world if we were not all trained as children to control our sphincter muscles. As adults, many harried, overstresssed people are learning various relaxation techniques that can be used to control blood pressure and other smooth-muscle conditions. Many of these techniques involve indirect control. Rather than willing his blood pressure to fall, a person may concentrate on a relaxing image, such as walking through the woods in fall. The extent and kind of control we can have over smooth muscles and glands is currently the subject of some controversy, both theoretical and practical.

The characteristics of striate and smooth muscles are given below. Striate muscle may be seen to be the typical agent and expression of consciousness, and thus skeletal movements can suggest a mind in control. The control of smooth muscles is generally unconscious, and so their action does not imply consciousness. However, since there are many exceptions to both of these rules, it is not always simple to draw a compelling inference about the role of consciousness in some action.

CONTROL OF MUSCLES

TYPE OF MUSCLE	TRADITIONAL NAME	OBSERV-ABILITY	TYPE OF CONTROL
Smooth	Involuntary	Invisible	Unconscious
Striate	Voluntary	Visible	Conscious

What about other animals? Do they all possess muscles? Are any of their muscles similar to our own? What do their muscles imply about consciousness?

A few animal groups, such as the protozoa and the sponges, have no muscles at all. Despite their lack of muscles, most of these animals still have some powers of contraction. For example, many protozoa have filaments that can contract. This contraction can change the shape of these single-celled creatures and enable them to move or eat.

The rest of the animals all possess specialized muscles. Many kinds of animals have only smooth muscles. These muscles are characteristically arranged in layers. Roundworms, which are often internal parasites, have only one lengthwise layer, and so their only behavior is a sidewise motion. Most other animals have both lengthwise and circular smooth muscles, which make more forms of action possible. Earthworms have both kinds of muscles, and can thus move from side to side or contract lengthwise. Circular and lengthwise smooth muscles are also commonly found around the interior gut and along large blood vessels in most animals, including earthworms and people.

Striate muscles are found mainly in certain mollusks like scallops and squids, in the insects and their allies, and in the vertebrates. In all of these groups, striate muscles are typically attached to something hard: a shell, an element of external skeleton, a bone. Thus, as in ourselves, their mode of action can be fast and easily seen: the scallop leaps, the damselfly's wings beat, the bear rears up. Striate muscles expend more energy than smooth muscles. The quick movements that striate muscles allow have a cost: higher food demands. The extravagantly active barn swallow must pay for its acrobatics by gorging itself with insects.

What do the muscles of animals imply about consciousness? In our own species, striate muscle can be under conscious control, but smooth muscles are basically under unconscious control. Yet in ourselves, conscious influence on smooth muscles is certainly possible. Moreover, we can sometimes perceive a smooth muscle movement, even if we can't control it, such as in a stomach cramp. Perhaps the best thing that might be said about other animals is that quicker action of the striate muscles may hint at consciousness, while the slower movement of the smooth muscles may not necessarily suggest mind.

Making inferences about the consciousness of animals simply on the analogy of their muscles to our own is not satisfactory; there is more to a beast than muscles. It is always better to use more than a single analogy in seeking mind. The more information we find about the scallop, damselfly, or bear, the more certain will be our interpretation.

Almost all muscle is subject to some kind of control by nerve cells. This is true not only for striate muscles but for the often "spontaneously" active smooth muscles. Muscles reflect neural activity. We shall look further into neurons to begin to learn more about the basis for consciousness in other animals.

How do the neurons of other animals differ from our own? How are these neurons assembled into nervous systems? Do different nervous systems work on radically different principles? What would different kinds of neurons and nervous systems imply about consciousness?

To start with, we should remember that there can be behavior without neurons or nervous systems. Plants provide one example, but there are many more. Consider the protozoa. Since they are single-celled, they can hardly have separate neurons or a nervous system. Yet the protozoa can engage in coordinated behavior. The slipper-shaped paramecium, for instance, is covered by short fibers called cilia. These cilia beat in unison and can propel the paramecium forward or backward. The bloblike amoeba's means of travel is different: it flows along a surface, waving pseudopods as it goes. Pseudopods are temporary "feet" that can capture food particles.

How can the movements of cilia or of pseudopods be controlled within a single cell that lacks neurons? The physiologists Naitoh and Eckert (1969) have linked the coordinated beating of the paramecium's cilia to ions and biopotentials. (This should sound familiar by now.) If we prod the front end of a paramecium, its cilia will reverse their direction of movement and the paramecium will shuttle away from us. Our prod causes a dimpling of the cell membrane of the paramecium, and the membrane responds by increasing its permeability to calcium ions. These ions flow inward and this causes the voltage across the membrane to drop. This drop spreads along the surface of the cell and leads the cilia to reverse their direction. Thus the paramecium will swim backward in response to a poke at its front end. If we prod the paramecium at its rear end, there is a change in permeability to potassium ions; these ions flow out and the cell voltage increases. This causes the cilia to beat more quickly and the paramecium swims forward to escape the irritant to its rear.

The escape response of the paramecium again illustrates the fundamental association of behavior with ions and biopotentials. As with plants, it is difficult to imagine what, if any, kind of consciousness might be found in single-celled organisms that are so physically remote from ourselves. One can perform the

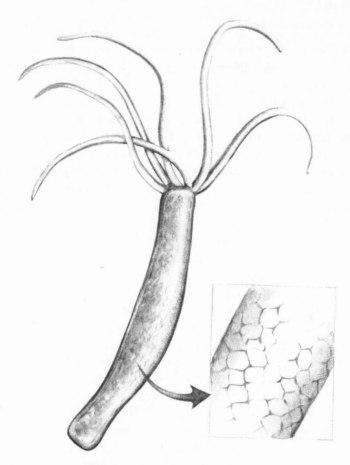

Hydra with enlarged nerve net.

exercise in imagination anyway, and try to think of the kinds of responses that a paramecium or an amoeba might make to different kinds of stimuli. What might it be like to be able only to discriminate a few classes of stimuli and to have only two or three basic kinds of responses? Margaret Washburn devoted a chapter of her classic text on animal mind to speculation along these lines and gave a tentative account of what an amoeba's mind might be like if indeed it had one. Following the scientific rule for economical explanation, her conclusion was that the mind must be simple, without reflection or memory.

Almost all multicellular animals have both neurons and a nervous system. The sponges, which are exceptions in many ways, lack both neurons and coordinated movements of the whole animal, although localized responses are possible. The simplest neurons and the simplest types of nervous system may be found among the jellyfish and its relatives in the coelenterates. Their neurons have a cell body and usually two or more simple axon-like extensions. Despite their lack of dendrites and their simple extensions, these neurons are essentially like our own: each neuron is a separate cell that is electrically excitable and that communicates through synapses with other neurons. Since it is possible for impulses to travel in either direction in many synapses, they are called unpolarized. The neurons are frequently joined together in one or two nerve nets. Impulses can spread in all directions in these networks. If two nerve nets are present in an animal, there may be no interconnection between them. It is as if there were two independent nervous systems in the same animal.

In certain corals and other colonial coelenterates, the nerve nets of individual animals are joined. That is, there is a direct neural connection that allows communication between colony members. This actual fusion of nervous systems is unique among animals. If ever there was a possibility for a group mind, this is it. Yet the nerve net is so unlike our own central nervous system that our limited powers of experience make it difficult for us to conceptualize the consciousness of a such a coelenterate. And imagine having such a nervous system linked to nervous systems of hundreds of others.

In any event, nerve nets are also prominent in roundworms, and in starfish and their allies. Nerve nets occur in other animals alongside other more important nerve structures. In our own species, nerve networks are found in the wall of the intestine.

Most animals feature two other types of nerve structure: *nerve cords* and *ganglia*. Nerve cords consist of bundles of nerve fibers. Ganglia consist of a number of nerve cell bodies along with connecting elements. Nerve cords and ganglia are

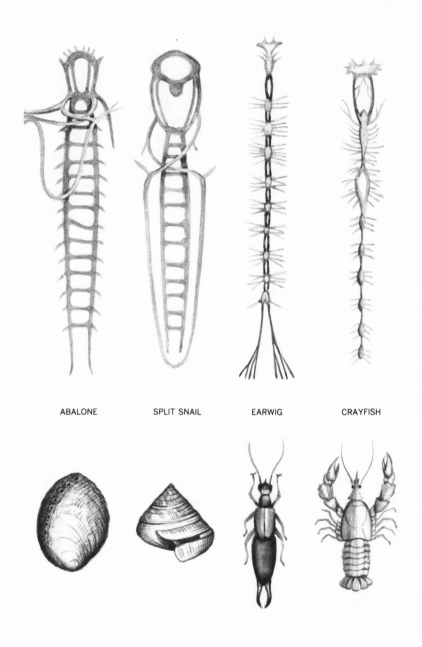

ABALONE   SPLIT SNAIL   EARWIG   CRAYFISH

Ladder nervous systems. Other forms of consciousness? The nervous systems of an abalone, a split snail, an earwig, and a crayfish are shown here. In some species, the two "sides" of the ladder tend to fuse.

the main structures of the common ladder type of nervous system. This kind of nervous system is found in groups as diverse as earthworms, insects, and mollusks, and is, in fact, characteristic of active mobile invertebrates. As may be seen in the illustration opposite, the head end of the basic ladder-type system consists of a ring of neurons around the animal's "throat." This ring has several ganglia, among which are the brain ganglia, located toward the back of the animal. Descending from the ring are two main nerves that form the sides of the ladder. Along these two nerves there are series of ganglia interconnected by transverse nerve cords, which form the rungs of the ladder.

The ladder type of nervous system is the most common one on earth and is arranged in a fashion that is clearly unlike our own. Since there are over a million insect species alone, there must be more than a million kinds of ladder nervous systems on our planet. If all invertebrates with ladder systems are conscious, then their sentience may also take more than a million different forms. To make such a possibility less abstract, let us consider two specific cases: the praying mantis and the octopus.

The male praying mantis's life is a dangerous one. In addition to all the predators he must escape in nature, he must also avoid being eaten by the larger female praying mantis. This insect femme fatale will eat anything that moves. To court the voracious female, the male must not let her "see" him. The male freezes when he first spies the female. He stalks her slowly, becoming still whenever she moves. When close enough, he hops on her back and begins to copulate. Unfortunately, the male may be spotted and eaten by the female before he can reach her. Even after mounting, the female may grasp and begin to consume him, head first. Despite this, the male continues to copulate. How is this headless mating possible? It turns out that the male's copulatory responses are inhibited by one of the ring ganglia on its head (Roeder 1967). As the female eats the male's head, she consumes the ganglion and totally disinhibits the male. The male can continue to copulate since this complex act is excited by the last ganglion in his abdomen, and this ganglion can still control behavior despite the loss of a head.

This cautionary tale of courtship can be taken in many ways. Let us consider its implications for consciousness. The praying mantis has a ladder-type nervous system, and in this system, different ganglia can control various complex acts. In fact, certain ganglia in the mantis can still continue to control acts despite the loss of a great part of the entire ladder system. If we think of consciousness as being involved in complex acts, then we must consider the possibility that different ganglia have independent consciousnesses, although there can be inter-relationships between ganglia. If a praying mantis has consciousness during complex acts, its consciousness might have a kind of multiple form that is quite unlike our own.

There is a radically different interpretation of the sacrifice of the praying mantis; its headless sex may mean that it has no consciousness at all. After all, if the mantis had no consciousness, there would be no reason to be surprised if part of its bodily machinery kept going after it lost its head. Thus it might be that the praying mantis is merely an unreflecting biological machine. Its nervous system may simply be a small, clever computer that senses nothing of the leafy green world through which it passes.

In their paper "The Insect Mind," J. L. and C. G. Gould argue that the behavior of insects is rigid and absolutely fixed by evolution. Because of this, they think that insects have no sentience. Along the same lines, they even express doubt that a great deal of human behavior suggests consciousness. There is no way to deny the Goulds' argument about insects; it may be quite right. Unfortunately, there is no way to confirm it either. None of us has shared the experiences of a mantis or a bee; we can't know its mind directly. Just because insects act mechanically doesn't necessarily mean that they are not conscious.

We are all qualified to comment on our own case. It certainly seems reasonable to think that many human acts are determined by evolution and caused by the unconscious. Yet even though we may not know the reasons why we act, we may

be quite conscious of the resulting experiences. Think how ri-
diculous human sex can be. Evolution and the unconscious fix
quite conscious desires in our hearts; we spend hours and hours
contriving to assume various odd postures. We don't know why
we want to get in these positions; the evolutionary reasons are
lost to us. Yet, despite the fact that much of our lovemaking is
mechanical, we are still exquisitely aware of what is happening.
So we cannot simply equate what is evolutionary or mechanical
with the unconscious.

The fact is that there is simply no parallel between the
brains of humans and the brains of praying mantises. While we
know that brains of our type can be conscious, we don't really
have any way of knowing about the radically different brains of
invertebrates. However, there is a strong parallel between the
elements of the two kinds of brains; the nerve cells in each
follow the same basic pattern. We can ask why, if nerve cells
underlie consciousness in our kind of brain, they should not
serve consciousness in another.

Consider the common octopus (*Octopus vulgaris*), a cousin
of the squid and cuttlefish and a more distant cousin of the snails
and clams. The octopus has been studied extensively, for it
survives well in aquaria and it has an intriguingly large brain.
In fact, its brain is on the same scale as that of a fish of equal size.
However, its brain does not look like the brain of a fish or a man.

Octopuses are thought to be quite clever by divers familiar
with them (e.g., Cousteau and Diolé 1973). Their lack of hard
body parts lets them find homes in all manner of crevices and
holes in the sea bottom; in one ancient Roman wreck in the
Mediterranean, they were found living in the ancient am-
phorae that had been scattered from the hulk. They often make
stone walls around their homes and they act like aquatic pack
rats, bringing coins and bits of pottery into their shelters. When
captured, octopuses are great escape artists; they can slither
and pry their way out of shipboard aquaria, using jet propulsion
to flee once back in the sea.

The reputation of octopuses does not suffer in the labora-

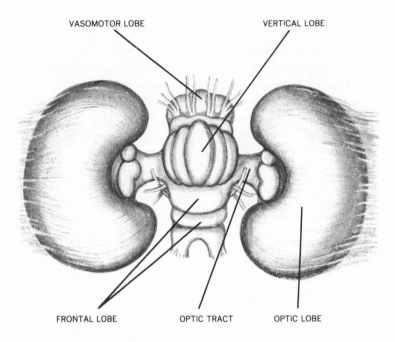

VASOMOTOR LOBE   VERTICAL LOBE

FRONTAL LOBE   OPTIC TRACT   OPTIC LOBE

View of the brain of the common octopus. Although some of the lobes are called by the same terms as in the vertebrates, the structures are not the same.

tory (Sanders 1975; Wells 1978). An octopus will investigate a new object placed in its aquarium by passing it along its arms to the beak. However, after a few exposures the octopus will either ignore the new stimulus or only briefly touch it with its arm tips. Octopuses can be trained to attack or pick up objects. They can learn to distinguish squares from diamonds, rectangles, and crosses, but not from circles. They show a form of "attention" in which they learn to respond to certain aspects of stimuli while ignoring others. All in all, a respectable academic performance for any animal, let alone an invertebrate.

Perhaps the most curious thing about the octopus is its many colors. An octopus normally closely mimics its background: it looks like the rocks, seaweed, sand, or even garbage it rests on. When disturbed, the octopus can literally change its spots. If one tries to force an octopus out of its shelter, the animal turns pale, and black circles form around its eyes. This has been taken as a sign of "anxiety." Following this, the octopus may draw its arm tips under its body and turn its suckers outward. The white-spotted sucker discs are set off by the color of the arm, which turns red-black. Another response to disturbance is called the cloud passage. Rings of color pass over the octopus's body, starting at the top of its head. Many human divers have been startled and confused by this display. When an octopus actually struggles, either with a diver or with another octopus, it may turn bright red and puff out its mantle, making it look much larger. To any animal, this eight-armed threat display might be quite intimidating.

These remarkable color changes result from the octopus's chromatophore and reflector cells. Hormones and nerves cause smooth muscles to expand the chromatophores, or color-producing cells. The expanded spots of color can make the octopus look brown, red, or yellow-orange. The reflector cells add blues, greens, or white light to the octopus's palette, making it a master of disguise as well as of display.

Many divers feel that the octopus's color changes hint at an emotional nature. Joanne Duffy interprets the responses of the giant octopus of the Pacific (*Octopus dofleini*) in this way:

> It is hard to keep in mind that octopuses of the size and weight of these are really very fragile animals, highly developed and with a very sensitive nervous system. They seem to succumb easily to nervous disorders. If a diver is too rough with an octopus, even without hurting it physically, it happens that the animal goes into a state of emotional shock and sometimes dies.
>
> When an octopus is angry, it often changes color. If, for example, you surprise an octopus on the bottom and touch it near the head, it turns white, and then brick-red—which is the color of discontent. It takes about two seconds for the entire body to change its color. Then the animal holds its arms over its head, like a protective helmet, and runs away in that position.

To sum up, we have many stories indicating that the octopus is a clever creature; in addition, we have hard scientific data that it can learn a great deal. The common octopus and some of its more mysterious relatives, like the giant squid, have the largest brains of any animals without backbones. In this, the octopuses and their cephalopod relatives are the primates of the invertebrates. What can we say about their consciousness?

Surely, an octopus is not a reflexive creature like an insect; yet its brain is much more like an insect's than our own. The brain of the octopus evolved from a ladder pattern, the same basic pattern shared by worms, insects, spiders, and clams. While many of these animals are rather inflexible, the octopus is not. In nature, the octopus is resourceful and has the capacity to change its responses through learning.

The octopus is perplexing. We might choose to deny insects sentience because they act mechanically, but how can we deny the sentience of the octopus? If an octopus is sentient, how can we interpret its consciousness? The form of the octopus's nervous system is so unlike ours—every lobe of its brain is different. Only the basic element of the nervous system, the neuron, is similar in form in both octopus and man. But what strong statement can we make about consciousness on the basis of neurons alone? And what are we to make of the rapid color changes that have suggested emotions to so many observers?

In 1973, Jacques-Yves Cousteau wrote:

> Cephalopods live in another world. I mean not only that they live in the sea, which is now open to exploration by human beings, but also that they inhabit a world of sensations and perceptions that is not our own. The evolutionary path which led cephalopods to a high degree of perfection is not that taken by the human race. It is nonetheless parallel to ours; and it may lead them further still.

The limit to cephalopod intelligence may lie in their green blood. The blood of cephalopods is based on copper compounds; this kind of blood is not nearly so efficient in absorbing oxygen as is the iron-based red blood of vertebrates. Cephalopods tire easily. Their brains cannot consume oxygen as rapidly as a vertebrate's. Thus, in its hundreds of millions of years of evolution in the sea, this alien intelligence may have always been constrained.

Reality is . . . a creation of the brain, a model of a possible world that makes sense of the mass of information that reaches us through our various sensory systems. All animals with sufficiently large brains create their various real worlds . . . although these worlds will differ one from another just as the brains that create them differ.
— H. J. JERISON,
Evolution of the
Brain and Intelligence

The brain is a machine for making analogical models.
— KENNETH CRAIK
*(quoted by
Warren S.
McCulloch)*

4

THE HOUSE OF CONSCIOUSNESS

What is our mind like?

Our mind is like our brain. Or should we say that our brain is like our mind? Perhaps what we mean is that our mind is a version of our brain. Or at least that our mind represents part of our brain, or a part of how our brain works. That much must be true.

What if there are brains that are like our own? Are there, then, minds that are like ours? This is the principle of analogy: like implies like. But does it always? How close must an analogy be before it becomes convincing? Are there brains so truly like our own that they can convince us that we share the earth with other forms of consciousness like ours?

Is our mind really like our brain? After all, it does seem odd that our mind should be located in a particular head on a certain blue planet that rides in the outwardly expanding mass of an exploded universe. And it may seem stranger still that our minds emerge out of a mist of atoms and spinning electrons that are ruled by the principles of chemistry and physics. There is a free, dimensionless quality of mind that suggests that we should not be attached to any particular planet or set of atoms. In many religions, consciousness is similarly conceived to be a free spirit that temporarily occupies a mortal body and that eventually will be released to wander in some form of infinity. A more prosaic view comes out of science, out of the observations and interpretations of the material world. This more earthly view was presented in the fifth century before Christ by the Greek physician Hippocrates. Writing about epilepsy, the sacred disease, Hippocrates taught that mind was in the head, in the brain, and not in the heart. Epilepsy itself was the result of brain abnormality and illustrated the link between mind and brain. Hippocrates wrote:

> Men ought to know that from the brain, and from the brain only, arise our pleasures, joys, laughter, and jests, as well as our sorrows, pains, griefs, and tears. Through it, in particular, we think, see, hear, and distinguish the bad from the good, the pleasant from the unpleasant.

Hippocrates' view of the location of mind prevails today. Most of us no longer follow the ancient belief that our minds are in our hearts or in our breathing. Instead, we think of our consciousness as being in our brains, about one-half inch behind our foreheads. Indeed, many of us think of our eyes as a pair of windows through which the mind looks out at the world.

Hippocrates' basic notion about epilepsy was also correct. Epilepsy often results from an abnormal irritation of brain tissue and does reveal a great deal about the relation of consciousness to the brain. This is so because epilepsy represents a blocking of normal brain activity that may or may not interfere with consciousness. This blocking may begin with abnormal activity at a limited site in the brain, although in some cases the abnormal discharge spreads even more widely. The sufferer from

Brain and spinal cord.

epilepsy may be conscious of the earlier and more limited discharge, but in the more massive storm of electrical activity, the epileptic loses consciousness.

If we want to look for consciousness in the brains of other animals, the study of epilepsy in ourselves can help us. Only we humans directly know that we are conscious, and only we can directly tie consciousness to the brain. In epilepsy, the link between consciousness and the brain is revealed in stark form. Before we proceed to learn about the place of the mind in the brain, however, we must know something about the basic layout of the brain. Although we all spend our existence in our brain, many of us are not familiar with this physical structure. So that we don't get lost, a floor plan is provided in the illustrations above and on page 50. Since the brain is the house of our experience, it is nice to know where its main rooms, stairways, and cellars may be found. After a brief tour we can try to relate these locations to consciousness.

The brain and its long extension, the spinal cord, are encased in the protective armor of skull and backbone. These

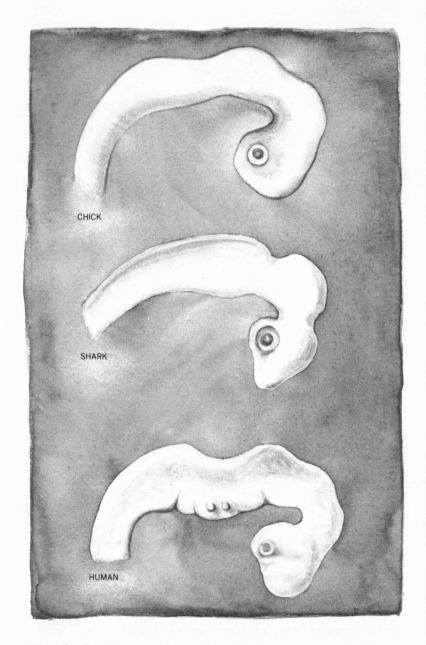

Neural tube. Where and when does consciousness begin? The neural tubes in the embryos of the chick, dogfish shark, and human are very similar.

fragile, protected structures constitute the central nervous system. Outside the bony armor are found the sensory receptors, the sensory and motor neurons, and the ganglia of the peripheral nervous system. The peripheral nervous system provides information about the external world to the brain and spinal cord and allows these central structures to control the muscles and glands of the body.

The brain and the spinal cord evolved from a simple neural tube. This neural tube is also present in our own embryos and becomes elaborated into our nervous systems. The "head" of the tube develops into the forebrain, midbrain, and hindbrain. The "tail" of the tube becomes the spinal cord. The center of the tube remains hollow and forms a series of "aquaducts" and "reservoirs" for cerebrospinal fluid. The structures in the forebrain area undergo the most differentiation as the embryo grows.

The forebrain develops into three structures. The first is the olfactory lobe, a sensory area for smell. The second is the pair of cerebral hemispheres that are so prominent in our species. These hemispheres come to form a caplike structure that covers the top of the brain. The hemispheres have an outer layer of gray cell bodies and an inner layer of axons that are surrounded by white myelin sheaths. The myelin sheaths permit fast neural conduction. The gray matter is called the *cortex*, from the Latin for "rind" or "bark." The hemispheres are divided into four major lobes: the frontal, parietal, temporal, and occipital. Hidden beneath this cap of gray and white matter is the third division of the original forebrain, the diencephalon. This contains a number of relay and control centers vital to such activities as eating and sleeping. Two important areas of the diencephalon are the *thalamus* and the *hypothalamus*. The egg-shaped thalamus lies above and behind the smaller hypothalamus, which is located at the bottom of the brain, right above the roof of our mouths. The thalamus makes direct connections to the cortex and the hypothalamus can release hormones into the bloodstream. Both of these avenues are important means of control for the brain.

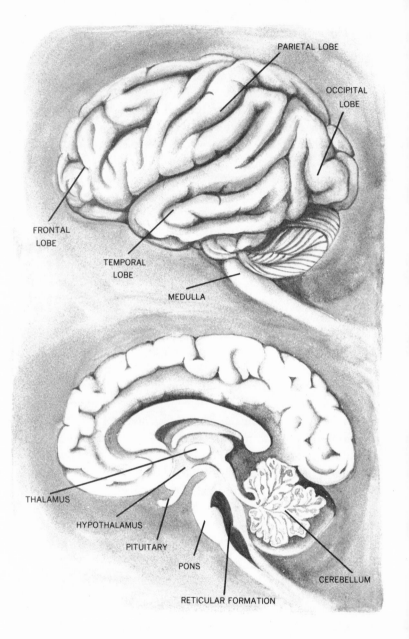

The house of consciousness. Side views of the brain. The surface of the left side of the brain is seen at the top. The structures within are revealed at the bottom.

The embryo midbrain becomes the tectum, which has a role in vision, as well as other centers. The hindbrain develops into three major structures: the pons, the cerebellum, and the medulla oblongata. The pons, which is Latin for bridge, links the cortex with the medulla and the cerebellum below. It also has sensory and motor centers for the head. The cerebellum, or "little brain," is concerned with balance and motor control. The medulla oblongata joins the brain to the spinal cord and is a control center for breathing and heart rhythm.

THE SOURCE OF CONSCIOUSNESS

Having finished this very brief tour of our nervous systems, let us return to the question of the site of consciousness. What areas of the brain participate in different kinds of experiences? What areas of the brain are absolutely necessary for consciousness? What does the study of epilepsy tell us about the site of consciousness?

Wilder Penfield, the late neurosurgeon from Montreal, was a pioneer in the study of consciousness and the brain. Penfield was a specialist who treated epilepsy by surgically removing the focus or source of epileptic discharge. To do this, he exposed the surface of the brain of a conscious patient after a local anesthetic had been applied to the incision. He then stimulated the brain with a "gentle" current of electricity. Since the brain itself has no feeling, the patient experienced no pain from the stimulation. This exploratory electrical stimulation was necessary to locate the source of the epileptic attack as well as to ensure that no critical areas were excised. Penfield's surgical technique could successfully eliminate severe epileptic attacks; it also revealed a wealth of information about the contribution of brain areas to consciousness.

One of Penfield's first discoveries was perhaps the most striking. In the course of his exploratory electrical stimulation

of the cortices of conscious patients, he unexpectedly entered
a storehouse of memory:

> On the first occasion, when one of these "flashbacks" was reported
> to me by a conscious patient . . . I was incredulous. On each
> subsequent occasion, I marveled. For example . . . a mother told
> me she was suddenly aware, as my electrode touched the cortex,
> of being in her kitchen listening to the voice of her little boy who
> was playing outside in the yard. She was aware of the neighbor-
> hood noises, such as passing motor cars, that might mean danger
> to him.
>
> A young man stated he was sitting at a baseball game in a
> small town and watching a little boy crawl under the fence to join
> the audience. Another was in a concert hall listening to music.
> "An orchestration," he explained. He could hear the different
> instruments [Penfield 1975].

Penfield had activated memories, and those memories
were recorded as if on a tape recorder. These reactivations of
the past were obtained when the temporal lobe was stimulated.

Stimulation in other cortical areas leads to different responses. Stimulation produces movements when it is applied to
the cortex that lies at the back of the frontal lobe. The movements produced from this motor cortex vary systematically
with the point of stimulation. Stimulation of the inner surfaces
of the cerebral hemispheres elicits foot movements. At the top
of the hemispheres, knee movements are made. As the electrode descends the side of the cortex, movements will be seen
in the arm, head, and tongue. Stimulation to the cortex at the
front of the parietal lobe produces bodily sensations. A map of
these points of stimulation reveals the distorted image of a
person, with enlarged areas for the hands, head, lips, and
tongue as may be seen in the illustration opposite.

Stimulation of the cortex with an electrode might be regarded as an artificial way of exploring the areas of the brain
involved in consciousness; in fact, identical patterns of response
are seen in focal epilepsy (Hopkins 1981). Focal epilepsy is a
condition in which a limited cortical area begins to show an
abnormal, chaotic discharge. The epileptic can perceive this

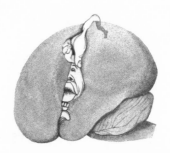

Musculus and Homunculus. At the surface of the cortex arrive impulses from the senses. The projections of the skin and body senses on the cortex take the form of the distorted figures of their owners—a "little man" and "little mouse" in this case. The most sensitive parts of the body—the mouse's whiskers in the top illustration, and the human's hands, lips, teeth, and tongue in the bottom illustration—have the greatest representation in these imaginative projections.

discharge as a sensation, a feeling, a thought, or a movement. In different cases, the sufferer may feel pins and needles in an arm, have a feeling of familiarity, or experience a complex hallucination. His mouth may twitch. Sometimes the patient is unable to speak, even though he is conscious. In other instances, the patient may continue some complex task, such as driving a car, yet later have no memory of the episode.

Focal epilepsy may remain limited to the original cortical area, or it may spread, culminating in a grand mal attack with both a loss of consciousness and convulsions. Grand mal attacks also occur as the immediate result of widespread abnormal discharge in the upper brainstem. These brainstem areas are not well defined; yet they are obviously necessary for the consciousness of epileptics as well as for the consciousness of those of us fortunate enough never to experience a seizure.

Medical evidence about the effects of drugs and injuries gives us additional information about the areas of the brain necessary for consciousness. One structure that does play a clear role in consciousness is the *brainstem reticular formation.*

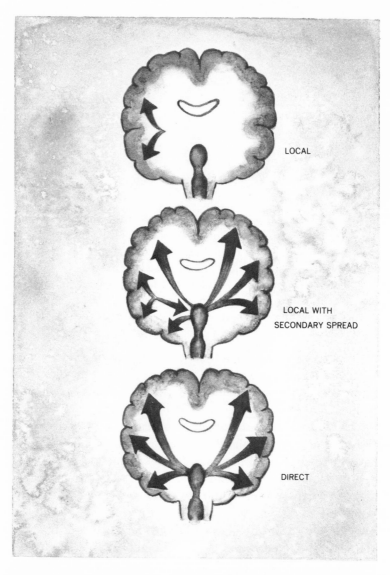

Epileptic seizures and the loss of consciousness. The "storm" of electrical activity that underlies seizures can remain localized (top) or spread widely in the brain if central areas become affected (center). These general storms can also originate directly in the central areas (bottom). Consciousness is lost when the abnormal discharges are widespread.

This finger-sized area of dense neurons lies at the core of the brainstem. Its short neurons make connections to incoming sensory neurons as well as to the cortex. Stimulation of the brainstem reticular formation makes animals look alert, as if they were watching for something. Injury to the reticular formation can render an animal drowsy or comatose. This formation would be well placed for a mechanism of attention or consciousness. According to one theory, the reticular formation can activate the cortex through its own direct upward connections. In turn, the cortex can moderate the level of reticular activity through its downward connections. Since the reticular formation is connected to incoming sensory neurons, it could also act as a gate for this information and control what actually reached the cortex. In the case of an epileptic attack, normal reticular activity would be disrupted by the electrical chaos of the grand mal seizure, and unconsciousness would result.

To summarize, research has begun to identify the structures that contribute to consciousness. We know that certain areas of the cortex help us control our muscles and process incoming sensory information. We know that other cortical and subcortical areas are concerned with memory. We know that brainstem core structures are absolutely necessary for consciousness. Impressive though these advances in our understanding may be, they also reveal our ignorance. We still don't know how the brain works as a whole. This means it is not always easy to interpret a particular fact. For example, if we stimulate a brain area and this makes a conscious patient mute, then we know that the area plays some role in speech. However, the interference we have created could be in memory, in perception, in muscle control, in thinking, or in anything else that might prevent speech. Likewise, if we see that abnormal electrical activity in the brainstem makes someone unconscious, we can safely deduce that normal brainstem functioning is necessary for consciousness. But we don't know how this normal functioning supports consciousness.

We must remember that at present the brain represents a mystery. Like good detectives, we can gather evidence about the brain and sometimes develop an interesting theory about how it works. But our case is still a thorny one. For example, there are too many suspects for the processes of consciousness, and no compelling way to choose among them. However, we do continue to find clues and get closer and closer to a solution to the mystery of the brain.

One possible answer to the question of the site of consciousness deserves mention. There is a suspicion that our very search for critical structures blinds us and keeps us from the truth. Suppose that consciousness does not occur in a few limited places in the brain, but is instead more widespread. Consider the possibility that consciousness may not be characteristic of a certain site, but rather of a certain kind of brain process.

What kind of process might be the ground of consciousness? One candidate has recently been proposed by Karl Pribram, who suspects that consciousness is based not on nerve impulses but rather on slower electrical voltage shifts. Such slow potential changes have been seen in layers of the cortex that have neurons with dense, intertwining dendrites. The slow potentials can be the result of volleys of brief nerve impulses arriving in parts of the dendritic network of different neurons. Patterns of arriving nerve impulses would change the local cell voltages slowly and produce the observed waxing and waning voltages.

Pribram's theory is consistent with several kinds of evidence. For example, our consciousness is not experienced in a staccato series of flashes that resemble a train of nerve impulses. Instead, our experiences seem to flow on a longer time scale, like that of the slow potentials. It also appears that our consciousness of events is not immediate but develops over time. When the body-sense cortex is stimulated with an electrode, as in Penfield's procedures, the patient will not report feeling anything until at least half a second has elapsed. It is as if the patient must await the generation of a slow potential to experience the stimulus, the initial nerve impulses being insufficient.

It is difficult to evaluate Pribram's theory, since we know so little about local slow potentials in the brain. Like other views about consciousness, the theory still needs hard evidence to back it up. Pribram's proposal should remind us, in any case, to keep an open mind about the physical basis of consciousness.

CONSCIOUSNESS IN OTHER BRAINS?

If we had the solution to the mystery of the physical nature of our own consciousness, we would be in a better position to search for consciousness elsewhere in the animal kingdom. If we found close parallels between ourselves and other animals in the structures and processes that underlie our consciousness, it would be convincing evidence that there is consciousness in other animals. Despite our ignorance about the precise source of consciousness, we can still begin to compare nervous systems. Is our brain like those of the other backboned animals? Is the cortex that literally crowns our nervous system unique to us? What differences can be found among the brains of our animal relations? What do these differences imply about consciousness?

Observe the brains depicted on the following page. From fish to mammal, the brain is formed on the same basic plan. From an embryonic forebrain comes the olfactory bulb, the cerebral hemispheres, and the diencephalon. Out of the midbrain develops the tectum. From the hindbrain emerges the cerebellum, and medulla oblongata. These terms should sound familiar, since they were all mentioned in our tour of the human brain: the fact is that the basking shark, the leopard frog, the Komodo dragon, the raven, and you and I all have brains that follow the same broad outline.

The general similarity extends to details in certain features. Almost all of our nerves enter the central nervous system either at the medulla or along the sides of the spinal cord. The cerebellum serves as a control for our coordination and has similar specialized cells throughout the vertebrates. The pituitary

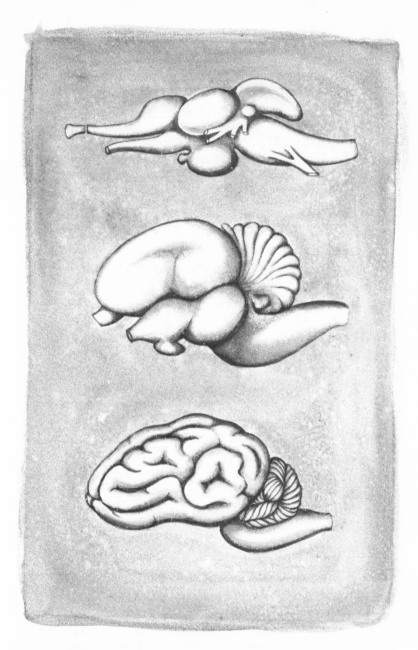

Vertebrate brains. (Based on an original illustration in Harry J. Jerrison 1969.) Top to bottom, brains of a fish, bird, and mammal are shown here. The brains are not drawn to scale.

gland, below the hypothalamus, secretes hormones in all of us. In these and other ways, there are many ancient channels through which information passes in all the vertebrate animals.

Though vertebrate brains are essentially similar, emphasis differs among species. In fish, amphibians, and reptiles, much of the information from the eye is put into action in the tectum, while this structure plays a lesser role in mammals. In fish and amphibians, there is little development of the cerebral hemispheres. In reptiles, the cortex appears; but in birds and mammals this outer bark of gray matter is highly developed.

What about the structures necessary to consciousness, those that lie at the core of the brainstem? These are conservative features in evolution: the brainstems of man and mackerel bear an essential similarity. The most striking features of the brain of the sheep, for example, are the cerebral hemispheres that are characteristic of birds and mammals. But if we look beneath these more recently evolved structures, we find the archaic brainstem, which resembles that of the early vertebrates.

To summarize this discussion of anatomy, we can say that the brains of all the backboned animals are alike in general outline. This similarity is even greater in the ancient structures that are necessary for consciousness in us. The obvious argument from analogy is that because of the essential structural similarity in our brains, all the vertebrates are conscious.

Can this argument be countered? It certainly can. We could say that we have not yet considered all the evidence from brain physiology or behavior. We could contend that we don't know enough about the anatomy of our own consciousness. True enough. Yet in one compelling way the evidence is plain. We can't dispute the broad plan of our anatomy; our brains are patterned like those of the other vertebrates. This can be taken as a strong argument that we all are conscious.

Still, many would not go along with the conclusion that fish or frogs are conscious, and would reject the basic analogy that can be drawn between our brains. They could argue that we and these cold-blooded, slimy creatures are different and so our brains are different too. Their brains are small and simple, while our brains are big and complex. Our large and complicated brains are the refuge of our dignity and our distinctive consciousness.

THE EVOLUTION OF THE BRAIN

There certainly is a great range in the size of brains: the brain of a minnow is the size of a match, while that of a sperm whale is the size of a basketball. Much of this difference is a matter of scale: the bigger the animal, the bigger the brain. However, there are differences far beyond those we would expect on the basis of body weight alone. Suppose we compared the size of different brains with the weight of their owners. On the graph opposite, the size of the brain is given on the vertical axis and the weight of the animal on the horizontal axis. As Harry Jerison of UCLA has noted, two patterns are easily seen in this figure. First, as we would expect, the heavier the animal, the greater the brain size. Second, some kinds of animals have brains that are relatively large or small for their weight. As may be seen, fish, amphibians, and reptiles have small brains for their body weight. Birds and mammals have brains that are relatively large for their body size. Primates (men, apes, and monkeys) and cetacea (whales, dolphins, porpoises) have brains that are the largest of all, relative to their body weight.

Jerison has looked at this relation in another way. Using body weight as a base, he computed the brain size that we would expect in a living mammal. He then compared those predicted sizes to those found in different kinds of animals. He included estimates for fossil specimens, which were calculated from the endocast, or inside impression, left by a fossil skull. Looking at brain size this way, we see dramatic differences between species. Brains of men and cetacea are several times the expected values for mammals, while those of fish, amphibi-

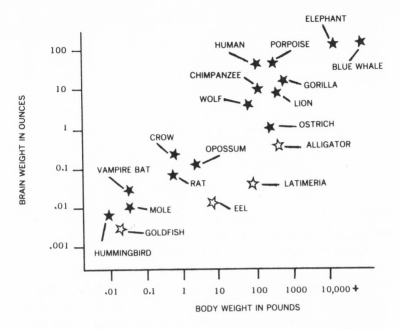

Brain size and body weight in the vertebrates. The bigger the body, the bigger the brain. However, the brain weights of birds and mammals (filled stars) are greater than would be expected by body weight, while the brain weights of fish, amphibians, and reptiles (open stars) are less.

ans, and reptiles are less than one-tenth the predicted size. Curiously, for many groups there does not appear to be any change in brain size over evolutionary time. The brains of fish, amphibians, and reptiles are the same size as they were when these groups originated. Archaic mammals, which seem to have lived a fugitive existence in the ages of reptiles and dinosaurs, had brains four times larger than their average reptilian counterparts. Yet for the 140 million years that reptiles ruled the earth, their brains stayed the same size. Following the disappearance of the dinosaurs some 65 million years ago, the brain sizes of the mammals began to increase. (Actually, some experts have recently reclassified the birds as specialized flying dinosaurs. If we accept their classification, then we would believe that small feathered dinosaurs are still around. In any event, the larger terrestrial dinosaurs abruptly left the face of the earth, possibly in the wintery aftermath of a collision of the earth with an asteroid. It is interesting to note that before their extinction, some exceptional dinosaurs had evolved brains larger than that of the average reptile.)

After a few skirmishes with some rather awesome terrestrial birds, mammals replaced the vanished ruling dinosaurs and came to dominate the land. Some mammals, such as the insectivores (moles, shrews, and hedgehogs), showed an initial increase in brain size, but then stopped, their brains today being of comparable size to those of 40 million years ago. Other mammals, such as carnivores and herbivores, continued to show an increase in brain size; the brains of today's wolf and deer are larger than those of their early ancestors. Primates also showed a trend for increasing brain size but were outpaced long ago by the cetacea. The brains of whales and dolphins reached their current great size some 20 million years ago. Mankind has only recently evolved its larger brain.

How does Jerison interpret these data? He feels that many of the increases in brain size were created by shifts in the senses used by evolving animals. When the archaic mammals were excluded from the daytime niches by the reptiles, their senses of smell and hearing became important to their new nocturnal existence. These senses required additional brain space for the ancient mammals to navigate effectively and locate objects in the dark. Once this adaptation was complete, there was no basic change in brain size until the end of the age of the dinosaurs. Once the mammals began to come out in the sun after the disappearance of the dinosaurs, vision became more important, and there was further development in the visual area of their brains. Some mammals remained at this stage of brain development; in others, brain size continued to grow with "intelligence."

After considering this evolutionary series of brains, can we say that the size of our brain clearly distinguishes us from other animals? No. Even the cortex so prominent in our species is not unique to us; our cortex is not radically different from those of other recently evolved mammals. Our great cortex does lie at one end of an evolutionary trend for increased brain size that began in the age of fishes, 400 million years ago. Yet we are not alone at our end of this evolutionary progression. As noted, our cortices were upstaged 20 million years ago by those of the cetacea. Long before our species evolved, certain cetaceans had brains of comparable size and complexity with those of contemporary humans. Most of that increased brain volume was in the very cortex that we humans have recently prided ourselves on.

If the relative size of the brain or cortex were the single criterion for consciousness, then some cetaceans should be close to our mental equals. In absolute size, the brains of many whales are far greater than our own. The killer whale, the gentle black and white *Orcinus orca,* has a brain that weighs 14 pounds (6.4 kg) compared to our three-pounder (the normal human range is from 2.8 to 3.7 pounds or 1.3 to 1.7 kg). The brain of the bottle-nosed dolphin (*Tursiops truncatus*) has been measured at 3.1 pounds (1.54 kg), similar in weight to our own.

While the brains of the cetacea may be rivals to ours in relative size, they are not exactly like ours in form. Some of the differences reflect changes in the shape of the skull over the course of evolution. Like us, the cetacea represent the extreme of a trend in evolution. Some 60 million years ago, the ancestors of the cetacea entered the sea. As they adapted to their new environment, their legs were transformed into flippers, their bodies became streamlined, and their sense of smell was diminished or lost. The shape of their brains also began to change. As we can see in the illustration on page 64, their cortex shortened and broadened until it became wider than it is long. The resulting rather globular shape of the cetacean brain contrasts with the more football-like brain of the other mammals. Within the domed cetacean brain, another difference developed, perhaps more significant than its shape. There is an extra cortical lobe that is unlike any possessed by the other animals. This paralimbic lobe is a center for incoming sensory information. We do not yet know its exact function.

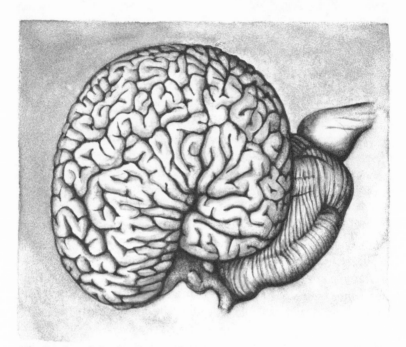

The brain of the bottle-nosed dolphin. The globelike brains of the cetaceans are not shaped like those of the other vertebrates.

The cetacean brain also differs from ours in less dramatic ways. There is a lower density of neurons in the cortex but a greater amount of cortical folding: the net result may be a similar number of cells. The motor areas of the cortex are smaller than expected, indicating that cetaceans no longer need to have as much coordination as their landbound kin; the sea buoys up the cetacea as it does fishes, and neither kind of animal requires all the complex behavior needed to fight gravity. There is relatively more sensory input from the head and the gut than in other mammals. The area of the brain devoted to hearing is large, reflecting the cetacean's use of echoes. Cetaceans make clicks through their blowholes and then analyze the echoes to orient and navigate in the sea. It has been suggested that the echo-sense of the cetaceans is so fine that dolphins and whales not only can detect each other but may even sense each other's guts. Like an obstetrician using a sonar system to form an image of an embryo, a cetacean may be able to "see" how another "feels."

The great cortical development we share with the cetaceans has led many to believe that we also share consciousness. Yet there are differences between our cortices—the cetacean's mysterious paralimbic lobe, its vast specialized area for hearing, and the lower density of cells in its cortical layers. These differences imply that our perception, our social communication, and the quality of our experience may all vary.

We should remember that we know far less about the cetacean brain than about our own. There are few specimens of many species, and study of living brains in large aquatic creatures is difficult, to say the least. Still, it is clear that the same anatomy that stands as evidence for similarity also stands as evidence for difference. Like implies like, and unlike implies unlike; in any less than perfect similarity, there is contrast. If we base our conclusions on a comparison of our brains, then cetacean consciousness should be both like and unlike ours.

Dolphin and manatee, wolf and sheep, lion and mouse—all the recently evolved mammals—share with us a prominent cortex. If a cortex distinguishes us, it also distinguishes our fellow recent mammals. Does this mean that we all share a distinctive mammalian form of consciousness? Would this consciousness be different from that of the archaic mammals of 100 million years ago? What about the consciousness of the reptiles that were ancestral to the archaic mammals?

What changes were created in the cortex during evolution? Ancient and contemporary reptiles had only the beginning of a cortex. Most of the behavior of a lizard or a turtle is controlled by the brainstem. When the archaic mammalian cortex appeared in the days of the dinosaurs, it formed a series of structures around the upper brainstem. This ancient cortex was not lost when the mammals replaced the dinosaurs. Instead, it remained beneath the new cortex, or neocortex, of the evolving modern mammals. Today we call this ancient cortex the limbic

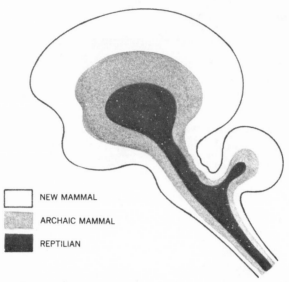

NEW MAMMAL

ARCHAIC MAMMAL

REPTILIAN

The triune brain of the mammals. (Based on an original illustration in Paul MacLean 1967.) The old cortex, or limbic system, of the archaic mammals developed out of the ancestral reptilian structures below it. In turn, the old cortex was surrounded by the new cortex as modern mammals evolved.

system. The limbic system lies near the center of our brain and of the brains of the other mammals. It is regarded as a center for emotions and for many other basic functions.

As Paul MacLean of the National Institutes of Mental Health has pointed out, the human brain contains its own past within itself. The ancestral reptilian core, the archaic mammalian cortex, and the neocortex from the last 60 million years are all represented in the brains of today's mammals:

A comparison of the brains of extant vertebrates together with an examination of the fossil record indicates that the human [cerebrum] evolved and expanded to its great size while retaining the features of the three basic evolutionary formations that reflect an ancestral relationship to reptiles, early mammals, and recent mammals. Radically different in structure and chemistry, and in an evolutionary sense countless generations apart, the three formations constitute a hierarchy of three brains in one—a triune brain [MacLean 1982, p. 291].

The fact that we possess these structures from the distant past has been interpreted to mean that we also retain traces of ancestral consciousness and behaviors. For example, Jason Brown (1977) believes that the original reptilian consciousness centered on the body, and did not experience objects. It is characterized by strong and direct drives, such as hunger and sex, that are expressed in a fixed set of responses. The evolution of the ancient cortex of the archaic animals introduced the experience of images and the investment of feelings in acts and images. This archaic consciousness is dreamlike and includes emotions like anger and fear. With the evolution of the neocortex, the mammals structured their experience into objects for the first time. These objects are outside the body in space and persist in time. In short, the brains of modern mammals create a reality that is similar in many ways to reality as we humans experience it.

Brown feels that human consciousness has reached further, in that much of human experience is structured by language. Despite our words, Brown believes that we humans still do experience all three earlier stages of consciousness. We still hunger, spend our nights in dreams, and conceive of the world as made up of objects in space. In fact, our return to such earlier forms of consciousness may be exaggerated if we are upset or psychologically troubled.

For MacLean and Brown, our brains are our link to the evolution of consciousness, just as our bodies are one link in a living chain that extends back at least a half-billion years. Their theories about the triple nature of behavior and experience are controversial, yet their underlying observations of the rings of ancient structures within our brains are undeniably true. In one way or another, our consciousness can not forget its past.

Mens agitat molem.
The mind moves matter.
—ANONYMOUS

5

THE MACHINERY OF EXPERIENCE

What is the mind like? It is like the brain. But is it like the whole brain?

What is the image on a television like? It is like the television, but not like the whole set. It is not like the shape of the picture tube or like the design of the circuitry or like the crystalline structure of the transistors. The image we see is like the flat and glowing screen. In a much more indirect way, the image is also like the stream of electrons that sweeps the screen from behind. But most watchers are conscious only of the glowing screen. They are not conscious of the rest of the set.

So it is with our brain. We are not conscious of the form of our brain. We are not conscious of the shape of the brainstem or of the six layers of the cortex or of the paths that the nerves follow as they bring us information from the outer world. We are not conscious of the individual nerve impulses within the neurons of the brain. But we are conscious of something in the brain, even though we cannot say as yet where or what it is.

How can we find what we are conscious of? In the last chapter we noted that consciousness may not depend so much on a particular site in the brain as on a particular kind of process within certain areas of the brain. In other words, even though certain brain structures may be necessary for consciousness, not all processes within these structures will come to consciousness. For example, as we read these words, we all need to have our brainstems working properly to maintain our consciousness. But we certainly are not conscious of all of the thousands of signals being processed in our brainstems at this very minute. At most, we are conscious of only a fraction of these signals, and then perhaps only indirectly.

Where do we begin our search for those processes that actually come to consciousness? There is a logical place to start: we can look for any distinctive physical events that go along with our own consciousness. If we find that there are distinguishing electrical or chemical events that accompany our consciousness, we can expect that these events are tied in some way to those underlying physical processes that we do directly experience. Even though we may not completely understand the relation of the events to consciousness, we will have begun to nail down the physical processes that immediately result in experience.

It is also possible to look at the brains of other animals to seek traces of consciousness. If another animal has brain waves or brain chemistry like those that we can tie to our own consciousness, this suggests that the animal may be conscious. It also means that we can study the brains of other animals to better understand our own consciousness. In fact, this is exactly what we do.

Enough tarrying over words, let us set out on the trail of consciousness. Remember that we are like detectives tracking an elusive and clever quarry. Let us go out into the mist and seek the evidence the mind leaves in its wake.

Although many behaviorists have believed the mind to be an invisible phantom, a ghost unattached to the material world, this isn't the case. Consciousness leaves its telltale marks, its earthly tracks. It is true that this trail may sometimes be hard to see; but it is there as surely as thought is there. What does the trail look like? What kind of signs should we search for? What are the clues that the mind drops as it passes through the brain?

Gadzooks! The first thing that we discover as we study the ground looking for the signs of mind is that there is not just one trail, but several. In passing, consciousness expresses itself in several physical ways. It may leave a permanent trace or show itself more fleetingly.

TRAILS OF CONSCIOUSNESS:
The Material Expressions of Mind

EXPRESSION	COMMENT
Structural	Permanent changes in brain structure following experience
Electrical	Voltages recorded from the brain relating to consciousness
Chemical	Chemicals that underlie our awareness
Behavioral	Behaviors that go with awareness

LASTING TRACES OF CONSCIOUSNESS IN THE BRAIN

For years, people have believed that experience leaves a trace in the structure of the brain. For example, many have thought that if we remember some event, there must be a record left in the connections between the neurons of our brain—in other words, that our nerve cells have been "rewired" by experience. Others have suggested that repeated experience can even change the size or proportion of brain structures. The problem with these views is that the suspected changes in structure have often proven difficult or impossible to find. Observations of brain structure are hard to make; moreover, there are a great many places to look for suspected changes. Given the vast number of sites and cells in the brain, looking for the traces of experience has often seemed like looking for the proverbial needle in the haystack. Nevertheless, a start has been made on this trail of mind. If one looks in the right place, structural changes can be detected. For instance, laboratory rats normally spend their lives in prisonlike tiers of cages. If, instead, rat pups are raised in a special "enriched" environment, a large cage with runs and ramps and other rats, their cortices become significantly larger than those of rats reared alone in plain cages.

Another case is that of day-old chicks learning about chicken feed. If these chicks must learn to distinguish between ordinary chicken feed and bitter indigestible beads, their brains will come to differ from those of chicks that just find feed pellets. Specifically, the more widely experienced chicks develop more synapses and these synapses contain more transmitters.

While the structural path of consciousness has been traced only rarely, other paths of mind are easier to follow. The most obvious is one we can see plainly with the naked eye: we watch how a person or animal behaves to tell whether it is conscious. Two other paths of mind require a few aids to become visible. With our detective's polygraph we can follow the electrical traces of consciousness. With some food and drink and a cabinet of other substances, we can begin to discover what chemicals are closely linked to consciousness.

In this chapter we will look at those changes in electrical activity that accompany awareness. In the following chapter we will examine those chemicals that enter experience. In the chapter following that, we will consider behavior as an expression of consciousness.

ELECTRICAL SIGNS OF MIND

First, let us turn to the electrical correlates of consciousness in ourselves. Then we will look for these signs of mind in the brains of other animals. To do this, we will need a few electronic instruments. These instruments are not difficult to understand, since they are rather like the radios or television sets that are all around us. In a radio or a television, an antenna picks up a weak broadcast signal. This faint signal is amplified and then transformed into a sound that we can hear or an image that we can see. When we look at the electrical activity of the brain, the same thing happens. We use an electrode like an antenna to pick up the weak signals the brain produces. These signals are amplified and then can be played over a speaker, displayed on a television screen, or even written out on a strip of paper.

ELECTRICAL SIGNS OF CONSCIOUSNESS

Brain waves	Small voltages at the scalp or cortex
Evoked potentials	Shifts in voltage when a stimulus is given
Spike potentials	Discharges in single cells

Brain Waves

The first of these brain broadcasts to be received was from the cortex of a rabbit in a pioneering study done by the English physiologist Richard Caton in 1875. Caton found that there was a small voltage at the surface of the cortex that was in "constant fluctuation." Following Caton, others found similar potentials on the skull or cortex of other animals. These potentials were on the order of millivolts and represented an external measure of the electrical activity within the brain. The study of the activity of the human brain began in 1924 when the secretive Austrian psychiatrist Hans Berger placed electrodes on the human scalp. When the output from these electrodes was amplified, Berger discovered that a very low voltage was present, and that this voltage fluctuated just as in the case of other animals. Today we familiarly call these fluctuations brain waves; when he finally revealed his work in 1929, Hans Berger called his record of voltage changes an electroencephalogram (or EEG).

It was not until the 1950s that a clear outline of brain waves was finally drawn. We know more about human brain waves than those of any other animal. When you and I are alert, our brain waves have a rapid, irregular form and are of relatively low voltage. If we begin to relax and daydream, a slower pattern of low-voltage waves emerges. These waves occur about ten times a second and are known as the alpha rhythm. If we become drowsy, then a higher-voltage theta rhythm may sometimes occur, with a frequency of 5 to 7 cycles per second. If we fall asleep, then our brain waves gradually become slower and higher. We pass through four stages of deepening sleep, and are harder to wake in each succeeding stage; collectively these stages are called *slow-wave sleep.* Finally, after about 60 to 90 minutes, our brain waves abruptly become fast, irregular, and shallow, just as if we were awake and alert. However, our skeletal muscles are virtually paralyzed, even though our hearts may race and our breathing may quicken. While only a few twitches may be seen in our bodies, our eyes begin a series of rapid movements. This rapid eye movement or *REM sleep* is the sleep of dreams; about 80 percent of the time, if we are awakened from REM sleep, we will report that we are dreaming. If no one awakens us, our first REM sleep ends after about ten minutes. We return to another cycle of slow-wave sleep, which is then followed by a second REM sleep. As the night progresses, the depth of slow-wave sleep cycles decreases while the length of the REM sleep increases. At the end of the night, an hour-long dream feature becomes possible.

Note that what we have done here is to describe several states of human consciousness. Each state is associated with a characteristic pattern of brain waves. We all experience alert consciousness and this is associated with fast, low-voltage brain waves. Alpha and theta waves can occur in more relaxed states.

Brain waves in humans. The pen of the polygraph records small voltages at the scalp and at the muscles of the neck. Fast, lower voltage brain waves can mean either wakefulness or dreaming. Slower and deeper brain waves are seen in ordinary sleep. Muscle tension is high during wakefulness but is reduced in slow-wave sleep. Dreaming is marked by a virtual paralysis of the skeletal muscles.

These rhythms can also be produced by Zen monks and others through meditation. Finally we have the two states of sleep and their slow or fast EEG patterns. There is some question as to whether the two kinds of sleep should be regarded as conscious or unconscious states. Some argue that the events of sleep belong to the unconscious, since we usually can't recall them. Despite this failure of memory, many of us feel that we have a form of consciousness as we sleep. In fact, REM sleep can be an intense kind of experience. I frequently dream of flying and I know that I take great pleasure in these flights, though I will not remember them unless I awaken from the dream. Even during slow-wave sleep, we seem to be experiencing something 60 percent of the time. This mental content tends to be more fragmentary and less fanciful than dreams and is often concerned with recent events (Dement 1978). Coma seems to be the only time during which a human does not experience something.

Do other animals have similar cycles of wakefulness and sleep? Just by looking, we can see that many animals have an alternating pattern of activity and rest that resembles our own. Lion and lamb, owl and swallow, all have times of action and times of quiet. What are their brain waves like during these periods of action and inaction? Consider an animal whose cycle of activity and rest is very familiar to us—the cat.

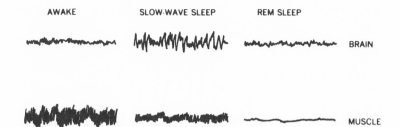

Brain waves in cats. Perchance to dream? The pattern of brain waves and muscle tension in the cat is strikingly similar to our own. Note that rapid eye movement sleep is marked by fast, low brain waves but little or no muscle tension.

Let us look at a polygraphic record of small voltages recorded from the cortex of a cat and from its neck and eye muscles. When the cat is alert, what we see in the record are fast, small voltage fluctuations at the surface of the brain; the voltages measured at the neck and eye show shifts as the cat moves its eyes and head. When our cat curls into a ball and goes to sleep, the record shows that although it maintains some tension in its neck, its eye movements diminish, and its brain waves become slower and deeper. After a few minutes of this type of sleep, a new pattern appears in the record. The brain waves become fast and low, as if the cat were awake, but the voltages recorded from the neck indicate that the cat is virtually paralyzed. This description of the brain waves of the cat should be recognizable; it is just like that of a human who passes from a waking state to quiet sleep and then into the sleep of dreams.

Additional studies have shown that the brain waves of all mammals and birds are basically alike. In both groups, active animals have fast, low-voltage brain waves. Inactive animals show either slow-wave or REM sleep. The only exception is the spiny anteater or echidna (*Tachyglossus aculeatus*) of Australia, which is one of the last living representatives of an early group of mammals. This odd egg-laying animal has only slow-wave sleep.

There are certainly differences in the details of sleep in different mammals and birds. Mammals, for example, vary in how deeply they sleep (Allison and Van Twyver 1970; Allison and Ciconetti 1976). Some are good sleepers; they sleep readily in a strange place, such as a laboratory, and will often sleep for more than eight hours. Other mammals have a more fitful rest; they tend to have a harder time adjusting to new surroundings and are likely to sleep fewer hours. Cats, dogs, and other predators tend to be good sleepers. Rabbits and other prey animals tend to be poor sleepers. However, any animal that sleeps in a secure place may sleep long and deeply. Bats in their caves are very deep sleepers, as are ground squirrels in their dens. In contrast, animals that rest in the open tend to sleep and dream less; sheep, for example, spend much of their night awake and dream only 6 percent of their sleep time. Baby animals have more REM sleep than adults. Lambs spend about 17 percent of their total sleep in a REM state, in contrast to the ewes just mentioned (Jouvet 1967, 1974). At birth, kittens have only REM sleep; only later does slow-wave sleep appear. Newborn human babies spend half of their sleep in the REM state; as adults, we dream away one-quarter of our sleep.

The sleep of birds does have some general differences from that of mammals. For example, the REM sleep of birds is shorter. The common pigeon has REM episodes that average only about seven seconds (Van Twyver and Allison 1972). Thus the city pigeon on its ledge has but fleeting dreams, at least by the standard of mammalian sleep.

The consistency of REM sleep in mammals and birds enables us to frame an answer to an ancient question. Does our dog dream as it twitches and whines in its sleep? Yes, it does, if we accept the evidence of the polygraph. The parallels seen in eye movements, muscle tone, and brain waves suggest that men and dogs are dreamers, just as the ancient Greeks thought.

If we are convinced that animals dream, then we can ask the question, What do they dream about? Freud wrote:

I do not myself know what animals dream of. But a proverb, to which my attention was drawn by one of my students, does claim to know. "What," asks the proverb, do geese dream of?" and it replies, "Of maize." The whole theory that dreams are wish fulfillments is contained in these two phrases [1927].

Whether Freud was right that animals dream of fulfilled wishes can't be shown to be true. But it is actually possible to form some idea of what animals dream about. Suppose the muscular paralysis of sleep is removed and the dreamer is freed to act out his dream? The block against action has been removed experimentally in cats, with the following results:

Although completely asleep, the cat will display behavior almost identical to that during the waking state. It will rise, walk about, attack invisible enemies, stalk an imaginary prey, or quietly sit and follow an unseen object with its eyes for a period of several minutes—all while deeply asleep [Allison and Van Twyver 1970].

It seems that cats, at least, dream of catlike things.

What can we say about the brain waves of the other vertebrates—of the fish, amphibians, and reptiles? Unfortunately, there has been only a single attempt to measure the brain waves of fish. This is understandable, since it is difficult to observe most fish in their natural setting, let alone record brain waves from free-swimming animals. Thus we really can't say much about the twenty thousand species of fish. What we know about amphibians and reptiles suggests that a variety of brain waves are possible. For example, tree frogs show a wave form of brain activity when they are moving about, but fast, low patterns when they are inactive, which is the opposite of mammals and birds (Hobson 1968). In contrast, the tiger salamander (*Ambystoma tigrinum*) shows no striking difference in brain waves between rest and activity, although orientations are associated with higher-voltage waves. Interestingly, the bullfrog (*Rana catesbiana*) never appears to sleep at all. Such inconsistencies, and the lack of studies of many kinds of amphibians and reptiles, make it hard to speculate about the early evolution of sleep and brain waves. More information would certainly help us in understanding the origin of consciousness.

Thus far, in our observations of brain waves, we have been content to let our polygraph run and let our human and animal subjects alone, free to be active or inactive. This way of following the path of consciousness has proven to be quite rewarding. We have seen that distinctive patterns of brain waves are correlated with different states of action. The close parallels in our observations of ourselves, of other mammals, and of birds indicate that we may share a similar form of consciousness. Considering how elusive the mind can be, this is a neat feat of detective work. We have, as it were, established some of the regular electrical habits of our quarry, consciousness. However, the air of mystery lingers. We still do not understand our suspect's modus operandi, the ways in which brain waves actually figure in consciousness. How are the different kinds of brain waves actually linked to sleep and waking states? How do the waves originate?

To begin to answer these questions, we will have to look at some other aspects of brain electrical activity. We will also have to go beyond observation and begin to stimulate an animal and observe its response. We can think of the stimulation as a question we are asking, and the response as how the animal answers us. As detectives we must question our witnesses to consciousness, even though most animals can answer us only indirectly.

Evoked Potentials

Suppose that we are recording the brain waves of a cat. We wonder what would happen to the waves if we made some noise or turned on a light. We snap our fingers and look at the record; we notice a change in the movements of the pens, although it is a little difficult to make out the response among all the ongoing squiggles. The pattern of waves that we have produced in the brain waves is called an *evoked potential.*

In order to see the pattern of response clearly, we often have to repeat our stimulus and then take the average of the

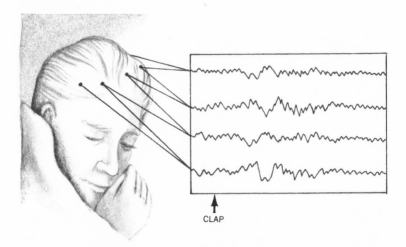

Human evoked potentials in a sleeping subject. (Based on an original illustration in Mary A. B. Brazier 1977.) The sound of clapping hands is reflected in brain waves recorded at several points of the scalp.

brain waves measured at that time. This is because ongoing brain waves are often so "noisy" that they can make the evoked response hard to see. When we average out the noise, we can see a plain pattern of evoked brain waves. This electrical response consists of a series of rises and falls in voltage, and it is centered in different areas of the brain. If we flash a light, then the waves will be first centered at the very back of the brain. If we make a sound, then the waves will be most intense at the very top of the cortex, in the area where the ear sends its nerve impulses.

The first peaks and troughs of the waves are seen whether or not the stimulus has any meaning for the animal. If the sound or light is meaningful, then later peaks in the waves can become prominent. For example, if we ask a human to tell us if he has heard a sound, then there is a distinctive later peak in the polygraph record. These changes can happen even if our subject is asleep. A positive peak can also be seen in a cat trained to respond to light or sounds (Wilder, Farley, and Starr 1981).

The computer has made the study of evoked brain potentials much easier. We are beginning to discover that certain types of brain waves can be linked to acts of perception or will

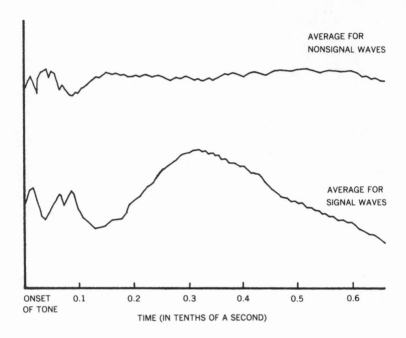

AVERAGE FOR NONSIGNAL WAVES

AVERAGE FOR SIGNAL WAVES

ONSET OF TONE

0.1 0.2 0.3 0.4 0.5 0.6

TIME (IN TENTHS OF A SECOND)

Evoked potentials in the cat. Once a cat learns that a tone is a signal, brain waves are evoked by this stimulus.

in people, and that parallels of these waves are sometimes apparent in animals. As we learn more, we may be better able to say what brain areas are involved in different experiences and how the evoked potentials of our brains and those of other animals compare.

Impulses from Single Cells: Messages from the Cat's Eye

Until now, we have been recording brain waves. The large electrodes we have used for our experiments have picked up the combined activity of many neurons. It is as if we were listening to a great crowd of people at a distance. We can hear the roar of the crowd but not the voice of any individual. Suppose that we want to listen to the individuals—that is, we want to hear what the individual neurons are saying. We can do this by placing an extremely small electrode in the brain. When we amplify the electrical activity recorded by such a microelectrode, we can hear or see the impulses of a single neuron within the brain. Played over a loudspeaker, the impulses make a popping sound. On polygraph paper, we can see a spike of voltage.

We can listen to the cells of the brain "going off" and try to relate their firing to the events of experience. This has been done only rarely in humans. In fact, it is the cat that was first studied intensively, and vision was the aspect of experience that was investigated. The story of vision begins in the back of the eye, with the special visual receptive cells—rods and cones. Oddly, these receptors face the back of the eye. Impulses are generated in the rods and the cones by light. These impulses are relayed up a chain of neurons to the back of the brain—the visual cortex in mammals. Axons arriving at this area behave like tiny targets for light. Each cell responds to a small round area in the eye's field of view. Light in the center of this area turns some cells on ("on cells") and other cells off ("off cells"). Light in the ring around this center "target" has the opposite effect; it turns on cells off, and off cells on.

In the visual cortex at the back of the brain, this incoming information is analyzed. The cells in this cortex are organized in columns. In each column are neurons that respond to more complex visual stimuli—slits, bars, and edges. The original inverted image of the eye is preserved at the cortex. The top of the visual field is represented at the bottom of the visual cortex and the bottom of the field is represented at the top of the cortex. The left half of the field is represented by the right hemisphere, and vice versa.

All of this sounds unfamiliar. Even though the primate visual system is constructed along the same lines as the cat's, we do not perceive the world as an inverted image of tiny spots, slits, bars, and edges. The visual cortex seems to fragment the visual world into a number of minuscule elements. In humans, at least, this process is unconscious; we do not see a grainy, distorted version of the outer world. Instead, we see an image from some further, yet-unexplored stage of cortical processing.

A single neuron firing in the cortex of a cat. When a small bar of light begins to fire, these discharges can be seen on an oscilloscope.

Other senses are similarly projected on the cortex in all mammals. For instance, each whisker on a mouse's snout is represented by a single barrel-like cluster of cortical cells. Yet the mouse may not be directly conscious of activity in these barrels, since consciousness might only occur at some more advanced stage of processing.

The question that lingers is where consciousness begins— both in the man and in the mouse. Researchers often joke about a legendary grandmother-detecting cell—a cell that stands for the unknown and further stage of brain processing where perception may actually begin. Even more difficult for us to find may be the cheese-detecting cell in mice. Where are those cells that directly underlie experience? We are coming to understand the machinery of the senses—but what about the machinery of thought? Much more patient detective work will be needed to find the hideout where consciousness abides.

We have come a long way since receiving the first broadcast from the brain of a rabbit. What we have seen is that in broad patterns the brain waves of mammals and birds are basically similar. However, the meaning of these patterns is not completely clear. The electrodes that receive waves from the brain are responding to signals that come from millions of cells; the waveform seen is the sum of the quick impulses and the slower voltage shifts in these cells. Thus, each wave must represent synchronous activity in vast numbers of neurons. But why this synchrony? And why is it so closely related to mind?

From looking at evoked potentials and single impulses, we have begun to locate and analyze processes in the brain. Yet the mystery of the electrical nature of mind remains. What is clear, however, is that man is not unique, not in the single impulses carried by his neurons or in the slower waves that sweep his cortex.

*The Master said, "The Path is not
far from man. When men try to
pursue a course, which is far
from the common indications of
consciousness, this course cannot
be considered the Path."*
—The Analects of Confucius

6

THE CHEMISTRY OF CONSCIOUSNESS

Each of us is a cauldron of chemicals, simmering at a steady body heat. Within the fifteen gallons or so in each vessel, there are trillions of chemical transformations. Consciousness arises out of such transformations; experience is the creation of chemistry.

Obviously, we are not conscious of all of the chemicals in the pot. Most chemical processes within us go unnoticed. After all, what consciousness could keep track of the chemistry of the billions of cells within our bodies? In fact, each cell of our bodies acts like an individual chemist running his own busy laboratory. Every cell takes in substances, breaks them down, and synthesizes new chemicals, all following its own code of instructions. How can we begin to find the chemicals that form the basis for consciousness in all this? Out of the innumerable chemicals created by our cells or ingested into our bodies, how can we isolate those that lie at the source of experience?

Let us ponder the matter in our detective's den. First, let us decant some sherry and light our pipes. Hmmm. What chemicals have a conscious effect? Aha, there's a clue—the fine Spanish sherry, with its slightly sweet and resinous taste. And there's another—the pipe and the aroma of the Carolina tobacco. The sherry and the tobacco certainly enter our consciousness. Initially we taste and smell the sweet residue of the grapes and the aromatic compounds of the tobacco leaves. Later we enjoy a pleasant kind of "rush" as alcohol, sugar, and nicotine pass into our bloodstreams and brains from the wine and smoke.

This armchair experience with wine and tobacco illustrates how closely mind is tied to matter; consciousness is tightly linked to chemistry. Indeed, we can think of a great catalog of substances that impress themselves on our experience as we smell, drink, or eat them.

What about the chemicals that originate within us? What chemicals produced by our bodies affect experience? Here our armchair method is not so helpful. We cannot sit comfortably in our dens, like Nero Wolfe sniffing orchids, and simply deduce the internal chemistry of mind. To find the chemicals that alter consciousness from within, we will need to turn to other animals for help. It was in their bodies that the internal chemicals underlying experience were first isolated. We will look at three kinds of these chemicals: the *transmitters,* the *hormones,* and the *opiates.*

THE CHEMICALS OF CONSCIOUSNESS

CHEMICAL	MAIN SOURCE	ACTION
Transmitters	Nervous system	Fast
Hormones	Glands	Slow
Internal opiates	Nervous system, gut	Fast or slow

THE HEART OF THE FROG

Two nerves go to the hearts of frogs. If one of these, the vagus nerve, is stimulated, the heart beats slower; if the other nerve is stimulated, the heart beats faster. In the 1920s, the Austrian physiologist Otto Loewi discovered that the neurons of the vagus nerve release a chemical where they touch the heart muscle. This chemical, *acetylcholine,* slows the beating of the heart as it passes across these synapses.

Acetylcholine thus became the first known *neurotransmitter,* a substance secreted by nerve cells to stimulate or inhibit target cells. Acetylcholine is not just a transmitter in frogs; it slows the beating of the heart in all the backboned animals, in cold-blooded fish and in warm-blooded men.

Acetylcholine is also released at many other internal organs besides the heart; it is the transmitter that the *parasympathetic nervous system* secretes at its junctions with smooth muscles and glands. As can be seen in the illustration below, the parasympathetic nervous system is one of the two branches of the *autonomic nervous system* of the backboned animals. The activity of the parasympathetic branch is greatest when we are calm and quiet. Our smooth muscles tend to be relaxed and our heart rate slows. Relatively greater quantities of acetylcholine are being secreted at the end synapses of our parasympathetic system. If instead we become angry or frightened, the *sympathetic nervous system* becomes more active. The sympathetic system is the second great branch of our autonomic nervous system. *Noradrenaline* is the transmitter secreted by the sympathetic branch at its synapses with smooth muscles, glands, and other tissues. When most of our smooth muscles begin to tense, when our hearts beat faster, and when our stomach secretions are reduced, we are feeling the force of our sympathetic nervous system.

The action of the sympathetic nervous system can also be illustrated by the frog's heart. If we stimulate the sympathetic

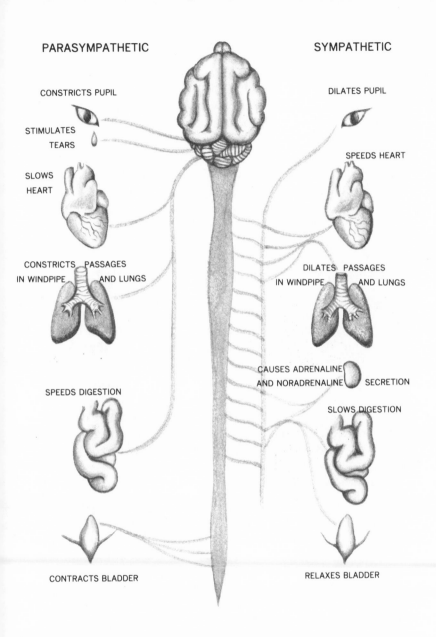

PARASYMPATHETIC

CONSTRICTS PUPIL

STIMULATES
TEARS

SLOWS
HEART

CONSTRICTS PASSAGES
IN WINDPIPE AND LUNGS

SPEEDS DIGESTION

CONTRACTS BLADDER

SYMPATHETIC

DILATES PUPIL

SPEEDS HEART

DILATES PASSAGES
IN WINDPIPE AND LUNGS

CAUSES ADRENALINE
AND NORADRENALINE SECRETION

SLOWS DIGESTION

RELAXES BLADDER

The autonomic system. Conscious or unconscious? The autonomic system in mammals regulates the stimulation of the cardiac and smooth muscles and of the glands.

nerve of a frog's heart, the heart beats faster. This is because noradrenaline has been released at the nerve-muscle synapses. Note that the action of the sympathetic system at the heart is the reverse of that of the parasympathetic system. The sympathetic system speeds the heart while the parasympathetic slows its beating. The sympathetic and parasympathetic systems of the vertebrates are often opposed in their actions. This duality in our autonomic nervous systems and presumably in our emotions is one of the great bonds shared by the backboned animals.

THE BRAIN OF THE RAT AND THE CAT

Neurotransmitters are found not only in the peripheral nervous system but also within the brain and the spinal cord. However, since there are billions of neurons within the brain, it is not easy to isolate and identify neurotransmitters in individual cells. Over the past thirty years a number of techniques have been devised to overcome this difficulty. For example, it has been discovered that some transmitters have a useful property: if they are exposed to the vapors of formaldehyde, they fluoresce green or yellow. In this way, we can accurately map the location of certain transmitters in the brain.

Much of the work of locating neurotransmitters has been done with the brains of rats and cats, but transmitters tend to be distributed in similar though not identical ways in the brains of many mammals. Let us look at a few transmitter maps and then relate these maps to consciousness.

At a brainstem structure called the *locus ceruleus*, or "blue place," are found many of the cell bodies of the neurons that release noradrenaline in the brain. The noradrenaline-secreting axons travel widely in the brain. One major group goes to the cortex and the cerebellum. Another group goes to the hypothalamus and nearby areas. The hypothalamus can activate the

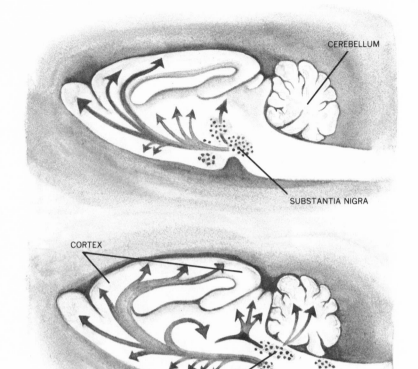

Dopamine and noradrenaline in the rat brain. Pathways of feeling? From cell bodies in the substantia nigra and nearby areas, dopamine-releasing axons travel "up" the brain to the frontal cortex and other targets. From cell bodies in the locus coeruleus and close-by centers, axons that distribute noradrenaline travel widely to many brain structures.

sympathetic nerves; thus noradrenaline is involved in the sympathetic system both inside and outside the brain.

Acetylcholine also serves as a transmitter within the brain. Acetylcholine-producing cells originate in or near the *basal nucleus*. Major projections go to the cortex and to the *hippocampus*. The hippocampus is necessary for the creation of

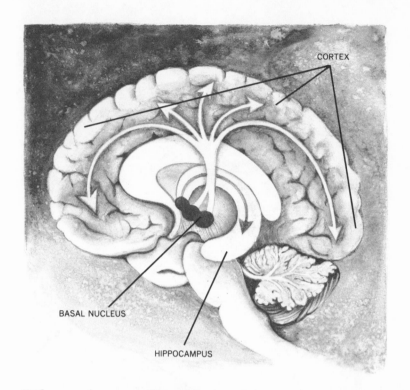

CORTEX

BASAL NUCLEUS

HIPPOCAMPUS

Pathways of memory. The cell bodies of the neurons that produce acetylcholine in the human brain are centered in and near the basal nucleus. The axons from these cells release acetylcholine in the hippocampus (dark arrow) and cortex (light arrows), structures important for memory.

new memories, and the cortex is involved in long-term memory storage. Damage to the basal nucleus and a reduction in acetylcholine production in humans occurs in Alzheimer's disease, in which there is a devastating loss of memory.

Another transmitter is called *dopamine.* The cell bodies of the neurons that release dopamine are mostly found in the *substantia nigra,* or "black substance," and in nearby areas. The axons from these cells travel upward to several places, including the ancient cortex at the base of the cerebral hemispheres. This ancient cortex is involved in muscle control. If dopamine is not secreted in this archaic structure, the body shakes with tremors. This is the basis of Parkinson's disease—a failure of dopamine secretion.

Several other transmitters have been studied intensively and many other substances are suspected to be transmitters in the brain; the list keeps getting longer.

CLASSIC NEUROTRANSMITTERS IN BACKBONED ANIMALS

TRANSMITTER	MAJOR ROLE
Noradrenaline	Arousal
Dopamine	Muscle control
Serotonin	Sleep
Acetylcholine	Memory
GABA	Inhibition

Any substance that serves as a transmitter must lie very close to consciousness. These chemicals are the media for communication in the brain; they spell the messages passed from neuron to neuron. The maps that are being made of transmitters may well prove to be maps of the mind.

How do we read these maps? How can we understand precisely what effect these transmitters have on consciousness? One way is to study how the chemicals affect behavior, both our own and that of other animals. A second way is to see how the chemicals affect experience; in this case the only direct evidence comes from human consciousness.

The most straightforward way to see how neurotransmitters affect behavior is to introduce them into the brain. This can be done if small tubes are permanently implanted in the brain of an animal. Through such tubes, minute doses of different substances can then be applied to very small areas of the brain. Following the introduction of the chemical, we can look for any changes in the way that the animal acts. For example, when the transmitter acetylcholine was injected in the side of the hypothalamus of a rat, the rat began to drink greedily. When noradrenaline was instead passed down the same fine tube, the rat would eat. This observation of a freely choosing rat suggests that at this site acetylcholine underlies thirst and noradrenaline underlies hunger.

Another example comes from the cat. At a site in the front of its hypothalamus, a small amount of the transmitter serotonin would lead to shivering and a decrease in body temperature. At the same site, a bit of noradrenaline would cause the cat's body temperature to rise.

The same chemical injected at different sites can produce very different effects. In the cat, injection of acetylcholine at different points can lead to circling movements, catatonic behavior (which is not catlike but trancelike), rage and attack, or purring.

Presumably these transmitters are involved in any conscious experiences that accompany these behaviors in the rat and the cat. In the human case, the experiences of hunger, thirst, chill, or heat might result from introductions of neurotransmitters like those given to the rat and cat. We do not know firsthand; we have not done these experiments on ourselves. Yet when we see a cat shiver or a rat eat, we can not but help to think that their chemical and mental processes may parallel our own.

THE DO-IT-YOURSELF CHEMISTRY OF EXPERIENCE

To see what effects different transmitters have in humans, we must alter our internal brain chemistry. This is easy to do; many of us change the state of the transmitters within our brains every day with our morning cups of coffee and our evening glasses of wine. In fact, most of the substances we ingest have at least an indirect effect on the transmitters in our brain. Most revealing to us, however, are those chemicals we take that either closely mimic or greatly interfere with natural transmitters. For example, many hallucinogenic drugs are structured like neurotransmitters. Mescaline, used in Native American religious ceremonies, is similar in its chemical form to noradrenaline and adrenaline. Taking mescaline leads to strong visual hallucinations. Presumably the source of these hallucinations is the overstimulation of the noradrenaline and / or adrenaline systems or sites within our brains.

Other drugs interfere with transmitters. Chlorpromazine (Thorazine) checks dopamine transmission because it ties up the receptors for the transmitter. The transmitting neurons can send dopamine, but the receiving neurons' receptive sites are occupied by the drug. This interference with dopamine transmission reduces many of the symptoms of schizophrenia. This and other evidence suggests that schizophrenia may be the consequence of an overactive dopamine system.

Let us take a much more familiar example of a drug influence on transmitters, that of coffee. It is thought that caffeine prevents the normal slowdown of the excitatory process that is started by neurotransmitters such as noradrenaline. If this theory is true, then those of us who depend on coffee to get started in the morning are experiencing a general rise in brain action from transmitters like noradrenaline.

All of this tells us that neurotransmitters direct what we do and what we feel. All these chemicals are very close to consciousness. While we certainly are far from understanding the entire chains of chemical causation that rule our experience, we have put our hands on some of the links.

THE COCK AND THE CAPON

Neurotransmitters are not the only internal chemicals that influence transmission in the brain. There is another class of chemicals that allows cells to communicate with others and that is tied closely to experience: the hormones. The power of this class of chemicals was convincingly revealed in a study by the

German zoologist Berthold in 1849. Berthold transplanted a testicle from a cock into a capon, causing the capon's comb to swell. This early example of hormone replacement demonstrated that reintroducing male sex hormones can create secondary male sexual characteristics. This experiment led to the monkey-gland craze, in which the testicles of monkeys were implanted in the bodies of men to increase their sex drive. The practice died out since such questionable implants do not survive, but the principle of hormonal action remains.

Since the time of the monkey-gland craze we have learned a great deal about hormones. Like neurotransmitters, hormones are released by specialized cells to influence target tissues. Although some hormones are produced by the nerve cells themselves, most are produced by our ductless glands: the pituitary, the testes, the ovaries, the adrenals, the pancreas, the thyroid, and the parathyroid. In addition, the liver, the stomach, and the duodenum secrete hormones. In contrast to the neurotransmitters, most hormones are released into the bloodstream. Thus their action tends to be slower and more widespread than that of the transmitters.

The action of many hormones is unconscious. For example, the hormone vasopressin is released by the pea-sized pituitary gland beneath our brain. This hormone lowers blood pressure. We humans are ordinarily unaware of the secretion of vasopressin and of its effects on blood pressure, even though hypertension can be a threat to life. We need to use a special instrument, a blood pressure cuff, to be aware of what our blood pressure is, and a much more complex chemical assay to find out about our vasopressin levels.

We seem to be largely unaware of the secretion and action of most of our hormones, to judge by the list given in the following table. Usually we are not conscious of what our thyroid and parathyroid glands are doing, nor are we aware of the actions of our liver, stomach, and intestines, except in the case of disease. As we can see, there are a great number of processes that hormones control unconsciously. This reminds us that much of what happens in our bodies is not revealed to consciousness.

Since we share all of the hormones in the following table with other vertebrates, we may suspect that they are also largely unaware of the secretions of many of their internal glands.

The hormones that are closest to consciousness are those of the adrenal medulla and sexual systems. Adrenal hormones flow when we are excited or stressed; the adrenal medulla secretes adrenaline and noradrenaline into the bloodstream and causes a whole series of emergency reactions.

Sex hormones become important to consciousness very early in life. Around the time of birth, the presence or absence of the male sexual hormone determines whether our sexual experience will be male or female. All brains in mammals begin life female; at a certain stage of growth, the testes of the male begin the secretion of testosterone. If the growing brain is bathed in this hormone, structural changes begin to take place; the brain assumes a characteristic male form. The time for these changes varies with the species. In the guinea pig, the brain changes its sexual form well before birth. In the rat, the brain assumes its sexual identity about the time of birth. In our own species, the brain of the unborn child takes the male form around the fifth month of life within the womb.

If testosterone is not available, the male mammal will be born with a female brain, despite its genes. The male's brain will remain female and its behavior will be feminized, although its body is still genetically male (see Durden-Smith and Desmond 1983). For example, when given estrogen, adult male rats that lacked exposure to testosterone around the time of birth will show the back-arching sexual posture of females.

In adolescence, the secretion of sex hormones increases. In male backboned animals, higher levels of testosterone are secreted by the testes. Secondary sexual characteristics appear. Maturing male warblers shed their drab fledgling coat of olive green to take on the bright reds, yellows, and blues of territory seekers: they start to sing the song of their kind. The young stag white-tailed deer puts its first horns into velvet and goes to rut. The young man grows a beard and speaks more deeply.

In the female vertebrate, the ancient reproductive cycle of

UNCONSCIOUS OR CONSCIOUS PROCESSES?
Important Hormones in Vertebrates

GLAND	HORMONE	TARGET AND FUNCTION
Hypothalamus	Releasing factors	Stimulates pituitary hormones
	Vasopressin	Regulates water retention by kidney
	Oxytocin	Contracts uterus; causes milk secretion
Pituitary	Tropic hormones, such as luteinizing hormone (LH)	Stimulates other endocrine glands like the testes and ovaries
	Growth hormone	Stimulates growth in all cells
	Prolactin	Stimulates milk production
Thyroid	Thyroxine	Regulates cell metabolism
	Thyrocalcitonin	Lowers calcium levels
Parathyroid	Parathormone	Regulates calcium and phosphate levels
Adrenal cortex	Steroid hormones	Regulate carbohydrates
	Androgens	Govern sexual characteristics
Adrenal medulla	Adrenaline, noradrenaline	General sympathetic activation
Liver	Somatomedin	Increases growth
Gut	Gastrin	Stimulates flow of gastric juices

UNCONSCIOUS OR CONSCIOUS PROCESSES? *(Continued)*
Important Hormones in Vertabrates

GLAND	HORMONE	TARGET AND FUNCTION
	Secretrin	Stimulates pancreatic secretion
	Enterogastrin	Inhibits release of gastric secretions
Ovary	Estrogen	Develops and maintains female sexual characteristics
	Progesterone	Regulates uterus
Testes	Testosterone	Develops and maintains male sexual characteristics
Pancreas	Insulin	Increases sugar storage
	Glucagon	Increases blood sugar

hormones commences. For example, in most breeds of sheep, the ewe lamb begins her cycle after midsummer. The shortening days cause the hypothalamus to secrete a releasing hormone. This hormone was first identified from samples of the brains of five million sheep (Guillemin and Burgus 1972). It was necessary to have this many samples since the quantity of the releasing hormone was so minute in any single hypothalamus. The releasing hormone travels over special blood vessels down the stalk of the pituitary. There it causes the release of the luteinizing hormone (LH) and the follicle-stimulating hormone (FSH). These hormones in turn flow through the bloodstream to the ovary. FSH causes the ovary to secrete estrogens, which act to establish and maintain the secondary sexual characteristics.

FSH also triggers the sequence of follicle growth. Once the egg is released, LH causes the formation of the corpus luteum, which secretes estrogen and progesterone. The hypothalamus reacts to the increased levels of estrogen and progesterone by cutting off the flow of the original releasing hormone. The lamb becomes a ewe in the waning days of summer.

Sex hormones are powerful influences on the brains and on the behaviors of animals like us. But how precisely do these hormones work? Do they directly fire neurons? Or do they merely activate synapses or circuits of neurons that are then fired by neurotransmitters? Or do testosterone, estrogen, and progesterone lead to both direct firing and indirect facilitation of nerve transmission? Or do they influence nerve growth itself?

One of the most remarkable studies in this light was the work of Fernando Nottebohm and his colleagues at Rockefeller University. Nottebohm showed that in spring the flow of testosterone in the male canary acts on a special center in its brain. The action of the hormone releases the species song of the canary. If a female canary is given testosterone, she too will develop a song center and begin to sing. Both male and treated females actually develop new neurons when they learn song (Kolata 1985).

As in the case of neurotransmitters, small amounts of hormones can be injected directly into the brain. The effects in the rat transcend sexual identity. When testosterone is injected into one of the areas of the hypothalamus, both male and female rats begin to make nests and retrieve young. If a second area of the hypothalamus is stimulated, male sexual behavior is the result —again in both male and female.

Whatever the mechanism, hormones certainly affect consciousness in the one species for which we have direct evidence. In woman, the monthly cycle of sex hormones can affect mood. In man, testosterone is important in the development of sexual interest. Since human hormone systems work on the ancient principles that have evolved with the backboned animals, we may suspect that there are parallels in the consciousness of other vertebrates as well. The ewe may experience changing moods as fall comes, and the ram a quite conscious interest in the signs of the ewe's mood.

SHERLOCK HOLMES AND THE CASE OF THE SNAIL'S FOOT

There is one final class of potent internal chemicals that must be mentioned. These chemicals were discovered only recently, despite the fact that a famous detective was personally entangled in the case over one hundred years ago. Sherlock Holmes abused both morphine and cocaine. Dr. Watson was outraged at Holmes's behavior:

> You know, too, what a black reaction comes upon you. Surely, the game is hardly worth the candle. Why should you, for a mere passing pleasure, risk the loss of those great powers with which you have been endowed?

Mr. Holmes replied:

> Give me problems, give me work, give me the most abstruse cryptogram, or the most intricate analysis, and I am in my own proper atmosphere. I can dispense then with artificial stimulants.

Odd it was then that Holmes never turned his logical powers to the mystery of his own pleasures. To try to solve the riddle of Holmes's addictions, let us begin with the source of many addictions—opium. Opium is obtained from the unripe pod of the opium poppy (*Papaver somniferum*). Opium and its derivatives morphine and codeine share some unusual properties. Extremely small quantities of these substances generate disproportionate degrees of pleasure. They also relieve pain dramatically; morphine is still the analgesic of choice on the battlefield. Close chemical relatives of the opiates, which vary merely in the symmetry of their molecules, will not produce pleasure; instead, they may block the pleasurable or pain-relieving effect of the opiates taken simultaneously. Such opium antagonists can be used to counteract overdoses or treat addiction.

Holmes might ask: How is it possible that opium could have so much power? The only other chemicals with such potency are neurotransmitters. And they too can be blocked with their mirror-image forms. What is so special about opium? In the 1970s, several investigators began to look at how opium worked in the brain (Davis 1984). First they mapped where opium acted in the nervous system. They found opiates acting exactly where one would expect: along the route that pain travels in the brain. But why should the brain have evolved special receptors for the milk of the unripe opium pod? Most known receptors within the brain are for the body's own chemicals—for neurotransmitters and hormones. Could there be some kind of pain-relieving chemical within the body?

Thus the next step was to find a substance with the properties of an opiate within the nervous system itself. In 1975 an internal opiate was found in the brains of pigs, guinea pigs, rats, and rabbits. It relieved pain, and its effects were blocked by an opium antagonist. Soon other internal opiates were discovered. The term *endorphin*, meaning "endogenous morphine," was coined for these internal chemicals.

Currently, three families of endorphins have been identified. These substances have their own pattern of distribution in the brain and are sometimes found in association with other kinds of brain transmitters. In addition, four types of opiate receptors have been identified. It is as if the nervous system were a potential palace of pleasures, our difficulty being that we don't quite know our way around its halls.

We do know that endorphins are released by a number of activities and events: running, birth, sex, and certain kinds of pain. During these events, there is often an experience of pleasure and a reduction in pain sensitivity. These effects can be abolished by opium antagonists.

Although originally it was thought that the internal opiates were possessed only by backboned animals, recently endorphins have been found in invertebrates. The terrestrial snail

Cepaea nemoralis will lift its foot if it touches a hot surface. This response is delayed by morphine but speeded up by an opiate antagonist. This indicates that this snail has internal opiate receptors that are stimulated by morphine but blocked by the antagonist (Kavaliers, Hirst, and Teskey 1983).

The fact that such diverse lifeforms as snails and men possess endorphins suggests that these chemicals originated long ago, when the ancestors of such contemporary groups first appeared. This means that the source of Mr. Holmes's addiction to morphine really lay in the evolution of endorphins in the late Proterozoic geological era, more than a half a billion years ago. Addiction to many other substances seems to trace back to the same primitive endorphin mechanisms.

What about our own experience with sherry and tobacco? Recent evidence is that alcohol is metabolized into an opiate-like chemical by the body. This externally derived opiate then acts on the delta opiate receptor. Tobacco acts differently but has the same end result: it releases natural endorphins that then bind the body's opiate receptors.

Transmitters, hormones, endorphins—the list of chemicals that act fairly directly on experience is growing. None of these chemicals is peculiarly human; we share them with many other kinds of animals. It would seem that the chemistry of consciousness is very ancient. Our feelings and our pleasures ultimately derive from chemicals first compounded in the warmth of primeval seas, when nervous systems began to evolve.

7

THE HIDDEN OTHER

The mind leaves many traces, as we have seen. Mysterious voltages ripple over the cortex, minute quantities of chemicals are released into synapses, interconnections between neurons are altered. All are signs of the mind's existence. Yet we do not need to be physiologists or chemists to find the track of consciousness; we have only to look at another being. When we see another person or animal behaving, we often can see the mind acted out before us.

Our self-knowledge tells us that much of our own behavior is the result of consciousness. Our acts reflect our thoughts, desires, wishes, passions. However, our behavior does not reveal the entire content of mind; that would be too embarrassing, at the least. Instead, our behavior expresses only a fraction of our consciousness. Indeed, most of us actively hide aspects of ourselves that could compromise or shame us.

We use our self-knowledge to infer consciousness in others. If other people look alert and generally act like ourselves, we assume they are conscious, too. But we go even further than that. We often think that we can tell something about the contents of their minds—their thoughts, their feelings, their motives. Again, we do this by looking for likenesses between us. If the other person responds like us in a similar situation, we tend to believe his or her experience is like ours. This is how we actually read minds; we reason by analogy to ourselves.

This business of inferring other minds is hardly foolproof. Often we cannot fully understand what the other person is responding to. If the woman across the room smiles, we may not know whether she is responding to a memory, an expectation, or a daydream. If we see a man doff his hat before some authority, does he act out of loyalty, fear, habit, or guile? The same act can have many meanings.

The unconscious further complicates our interpretation of ourselves and others. We are unaware of many of the things that we ourselves do. We can perform skilled, intelligent acts and be unconscious of the whole business; we can drive cars or balance on bicycles and be thinking about dinner or the boss. We can be completely unaware of the reasons for our own acts; we may dislike stringbeans or long skirts because of events long forgotten in childhood. Sometimes we even seem unaware of our own feelings; we can act aggressively and never perceive our anger toward child or mate. Sometimes we can see things in others that are invisible to them. Yet they may not recognize our interpretations of their behavior because they are oblivious to these motives or feelings.

Despite all the difficulties in inferring mind, the only way we reveal our consciousness is through our behavior. The sole exit for our minds, trapped within our bodies, is through our acts. Likewise, the only signs of consciousness we receive from others come through their actions. Behavior is the imperfect medium through which we show ourselves to one another. To know the truth about each other, we can only look at what we do.

There is one exception to this communication of consciousness through behavior: those cases where we measure brain chemistry or electrical signals directly, just as described in the last two chapters. In a sense, we are eavesdropping on the mind as it works. Yet the way the mind communicates is not well understood.

If our behavior is the only means by which we normally can reveal our consciousness, what about the behavior of animals? Do animals also act out aspects of their consciousness? Are the acts of animals windows to their minds? Or is the behavior of animals controlled by unconscious processes? Can the acts of animals tell us whether they are conscious or unconscious?

I look at the white cloud mountain fish (*Tanichthys albonubes*) swimming in the bowl on my desk. Their graceful movements are made by the propulsion of their pectoral, dorsal, and tail fins. They form a loose school and dart suddenly when I drop their food in. Within their inch-long bodies are nervous systems that are on the scale of a pin. These fish are so unlike me—I have no fins, no gills, no bright red spot at the base of my tail—yet their brains are formed on the same underlying pattern as mine. What acts of these fish can suggest to me that they are conscious?

To tell what acts suggest consciousness, we must first examine ourselves. We must look for those things in our behavior that show that we are conscious. Then we can search for the same things in the behavior of animals. In other words, we look for the signs of our consciousness in other animals.

What are the signs of consciousness? How do we test for consciousness in our own kind? Do you remember sitting in a grade-school classroom on a warm afternoon in spring? How would you tell which students were most conscious of what was happening? First, you might note whether a boy slumped over a desk reacted to anything—a frown, a joke, a spitball. If he was totally unresponsive, he might well be in deep sleep, a state that may not have a conscious content. Second, even if a boy seemed unaware of the blackboard and the teacher's voice, he might be looking out the window or staring into space. Staring may not always be a sign of consciousness, but in this case we might assume that a student gazing into some faraway scene is conscious, either daydreaming or wishing he were somewhere else. Third, we might see a student doodling or reading or playing. People who are being playful certainly are conscious, in my experience and probably in yours. Fourth, we may see that a handful of students are actually paying attention to the teacher and learning; they must be conscious of what the old dreadnought is saying. Finally, we can hear the students talk, in whispers or in response to the teacher's questions. For many, speech is the ultimate human sign of consciousness.

To generalize, these elementary tests of consciousness can be organized so that they apply to lifeforms other than ourselves. The first level of consciousness is sensitivity, which means simply that an animal reacts to a stimulus. The second level is that of orientation, which means that the animal directs its body in relation to the stimulus. The third level is that of exploration. This is a general term for all those investigations of new things, or of play. The fourth level is learning; the animal must remember something new. The fifth level is that of language, or of symbol and syntax.

In summary, each level represents a form of our own behavior when we are conscious. Because of this, when a human shows any of these acts, we tend to believe that he or she is aware. Each level is thus a sign of consciousness. If we see one of these signs in another animal, it may be taken to suggest that the animal may be conscious. Of course, by itself, each sign is not an infallible guide to consciousness, even in our species. We must weigh the evidence of all the signs.

THE SIGNS OF CONSCIOUSNESS

Sensitivity

The most simple act that reveals consciousness to us is a reaction to some stimulus. If we call someone's name and he or she looks up, we believe the person is conscious. Indeed, students of first aid are taught to ask a question like "Can you hear me?" to test for consciousness before rendering aid; it is both embarrassing and potentially dangerous to try to resuscitate someone who is not unconscious.

Sensitivity to stimuli is a universal trait in living beings. Bacteria respond to warmth, viruses to cold, trees to light, fish to vibrations—all living creatures respond to some kinds of stimuli. Their response may not be like ours and it may not be on our time scale, but it certainly represents sensitivity. Yet the dissimilarity of response between ourselves and other creatures may suggest either that consciousness in these lifeforms is different from ours, or that consciousness may not exist. For instance, a virus responds to a healthy cell by entering it and capturing its genetic machinery. Many viruses prefer cells at temperature of 85 to 92 degrees Fahrenheit, like those found in throats and mouths. Most of us do not go around invading cells, no matter what temperature it is, and would not accept such a response as compelling evidence for consciousness in viruses.

On the other hand, if we consider an animal whose responses are more like our own, the inference of consciousness may be stronger. For example, in heavy rain, sheep will turn to face out of the wind. Many humans trapped out in a downpour will also turn away from the wind. Seeing sheep in a storm reminds us of ourselves, and so we may suspect that sheep sense the rain and wind.

Orientation

We have previously noted that many plants turn toward light and away from gravity. A more general term for such tropisms is orientation, meaning that a lifeform orients its body in relation to some stimulus. Orientation is one of the commonest forms of behavior. Like sensitivity, it is found widely throughout the animal and plant kingdoms. Humans, too, show orientations; for example, much of our day is spent fighting gravity by maintaining an upright posture. We share our struggle against gravity with such diverse creatures as birds and trees. Yet while a cliff swallow's upright stance might suggest it was conscious of gravity, an oak tree's foursquare stance would hardly be so convincing. While the oak stands erect as we do, it is unlike us in other ways. In the oak there is nothing like our brain or our freedom of movement. If a lifeform acts like us in a single way but is unlike us in many other ways, it is hard to argue that it must be conscious.

Exploration

In exploration, a lifeform investigates something new in its environment. Exploration of the world is not seen in as many lifeforms as sensitivity or orientation. Trees do not explore their environments, but many mammals and birds do. Jackdaws, which are close relatives of the crows, are notoriously curious, and will investigate all kinds of new objects.

Exploration may turn into play. If tigers, leopards, or lions in zoos are given a ball or bit of rope, they will act just like kittens. Captive mice will sniff and play tug-of-war with rubber bands held at the side of their cages.

Exploration entails an animal taking in information about new things. In human experience, this is a time of intense conscious concentration on some new sound, image, or symbol. Our exploration can be directed toward other people as well as other things. As in other animals, our exploration shades off into play.

Another reason for including exploration in this list of tests for consciousness is that it entails all the electrical signs of physiological arousal. For example, there are low, fast brain waves, changing pulse rates, and blood flow to the brain when an animal approaches a new object.

Exploration is a very ancient and widespread behavior in vertebrates. For example, it is easy to observe in many fish. Many tropical fish respond to a gentle tapping or movement of fingers by stopping what they are doing, turning to the stimulus, and often approaching. Exploration is especially dramatic in the slow and stately movements of the South American angelfish (*Pterophyllum scalare*).

Learning

Learning was the traditional test of consciousness for the followers of Darwin, because it was closely linked to consciousness in human experience. This does seem to be true for some kinds of learning; we must be conscious in order to learn a new word in Chinese or how to play a new video game.

In contrast to such cases of learning, we know that we can react or orient to something without being conscious of it. As we walk, we may react to literally thousands of stimuli; however, we may be thinking of the moon or of the *Mona Lisa*'s smile. Thus, sensitivity and orientation are not always conscious.

The possibility that animals may not be conscious worried early thinkers about animal mind. Since sensitivity and orientation were possible without consciousness in ourselves, these characteristics were not considered convincing evidence for consciousness in other animals. Yet if learning occurred, there must be memory, and memory was surely proof of consciousness.

Actually, there is a flaw in this argument that was recognized long ago. Certain kinds of learning do seem to take place without awareness. Our stomachs learn to "expect" certain kinds of foods at certain times, yet we do not consider our stomachs to be conscious. Much of our unconscious learning seems to involve our internal organs; we often may become aware of the learning if we deviate from our routine. Fly to Tahiti tomorrow, and your jet lag will tell you that your body has indeed learned to function on the time back home.

On the other hand, it seems that we humans must be conscious during other kinds of learning. We must be awake to learn physical skills or to acquire new information. In sleep, which entails conscious elements, learning is minimal if it occurs at all. Sleep is distinguished by forgetting rather than by learning.

Let us look at three forms of animal learning to see which might entail consciousness. In the first, we condition a reflex. In the second, we create a new response in the animal. In the third, we teach the animal a "cognition," which is something like an inner representation of an outer fact or relationship.

REFLEX CONDITIONING.   In some kinds of learning, an animal attaches an existing response to a new stimulus. Ivan Pavlov, the Russian physiologist, was a pioneer in describing this kind of learning. In one of his experiments, a bell would sound and then a hungry dog would be given food. At first, the dog would only salivate when it was given the food. But after several repetitions of the bell-and-food sequence, the dog would come to drool when it heard the bell. Thus an old response, drooling, is given to a new stimulus, the bell. This kind of learning is hardly alien to us human beings; our mouths can begin to water at the sights or sounds of kitchens or even when we read a cookbook.

Whether consciousness is entailed in such conditioning, or to what extent, is questionable. We don't seem to be aware of learning to salivate at the sights or sounds that go along with food. Also, the kinds of response in this conditioning are primitive, reflexive, and internal, and involve the action of glands and smooth muscles. All of this may suggest that much of this conditioning is unconscious. Yet Pavlov certainly did not think so; he thought his dogs were aware. In fact, he used his experiments to investigate how they analyzed their environment.

The kind of learning studied by Pavlov is often called classical conditioning, since it was one of the first kinds of conditioning to be studied. Reflex conditioning is widespread in the animal kingdom (Corning, Dyal, and Willows 1973–75), but is not
found in plants. If we can get an animal to give a response, we
can often train the animal to respond to a new stimulus. For
example, there have been scores of studies of classical conditioning in planarian worms, which have simple nervous systems. These curious flatworms regenerate if they are cut in two.
If an "educated" flatworm is sliced in two, experiments indicate
that one or both regenerated halves may retain some of the
original conditioning.

Although classical conditioning may occur in flatworms,
snails, and octopuses, this may not imply that these animals are
conscious. We are not sure whether consciousness is necessary
for this kind of learning. Nor are we sure whether what is true
for us is true for animals with radically different brains from our
own. Until we fully understand the relation of consciousness to
conditioning in ourselves, we will find it hard to interpret conditioning in other lifeforms.

SHAPING A NEW RESPONSE.   Suppose we want to teach an animal a new response, one we rarely or never see. Let's say that
we want to train our dog to roll over. However, our dog does
not often roll over spontaneously; it would be peculiar if he did.
To make this rare or nonexistent behavior happen frequently,
we will have to shape the response. For this purpose we can use
some small treat or other reward. When we see some response
that even vaguely resembles rolling over, such as lying down,
we say, "Roll over," and give our reward despite Rover's sloppy
performance. Soon we find Rover lying down when he hears us
saying roll over. Now we start to shape Rover's response closer
to rolling. We wait until Rover is lying down but give the reward only when he begins to move in the direction opposite to
the one he lay down in. Once he's doing that, we then limit our
rewards to when he starts to regain his feet. By this process of

successive approximation we shape Rover's behavior into the form that we want.

Rolling over is an external response that is easy for us to see. This is so since it is made with the voluntary muscles, the muscles that make the skeleton move. Rover's new trick is different from the gastric secretions that were studied by Pavlov. We haven't merely taken an internal, reflexive act and conditioned it to a bell or buzzer. Instead, we have shaped the dog's "spontaneous" behavior into something new. In human terms, it's like learning a new skill. For most of us, we have to be alert to learn to hit a baseball or to turn a somersault; thus, if an animal learns a new response like rolling over, it is not unreasonable to think that consciousness was necessary.

The Harvard psychologist B. F. Skinner developed the technique of shaping responses in his work with pigeons and rats. Keller and Marian Breland, who were students of Skinner, were the first to take shaping and other training techniques out of the laboratory and into the field. The Brelands began an animal-training business with such creatures as "Priscilla, the Fastidious Pig," who turned on a radio, ate breakfast at a table, picked up her dirty clothes, and ran a vacuum cleaner. The Brelands became a financial success and their techniques have been used with such diverse creatures as opossums, porpoises, and whales.

If the shaping of a new response suggests consciousness, then it is significant that this training technique does not always succeed. New responses can be shaped in backboned animals, but seemingly only in a few invertebrates, such as the octopus. Among the vertebrates, certain mammals such as rats, pigs, and raccoons are especially versatile. Animals that can easily be shaped tend to have relatively large brains and to be either predators or omnivores.

COGNITIVE LEARNING. In the 1930s the star of behaviorism was in the ascendant. Many psychologists refused to use mentalistic terms and talked strictly of responses. At Berkeley, Edward

Chase Tolman dissented from this mindless psychology. Instead, Tolman believed that it was the very responses of rats in mazes that showed they had purposes and expectations; these mental processes determined what the rat did. Tolman devised many ingenious experiments to illustrate the quality of such cognitions in the rat. Cognition is a term that was adopted by Tolman as well as by more contemporary behaviorists to refer to mind. A cognition is some mental process, a "picture," an "expectancy," or even an "idea."

The "cognitive map" was one of Tolman's own cognitions about rats. Tolman thought that rats did not learn responses in mazes; instead they learned a map. If a rat was placed in a maze, it would wander. As it wandered, it would build up a map of the maze. If the rat later found food in the maze, it would be able to return to the food by using its map. Time has supported Tolman; there seems to be little question that rats create inner maps and follow them whenever it becomes advantageous.

Many other kinds of animals can form cognitive maps. For example, the beewolf, a solitary digger wasp (*Philanthus triangulum*), makes its nests in sandy ground in Holland. After a few hours spent digging its nest, the beewolf makes a circling flight near the nest entrance. During this flight, the beewolf learns the local landmarks that enable it to return to the nest after hunting. If the observer shifts prominent landmarks, such as pinecones, to a new location, the wasp will try to find its nest in the new and incorrect location (Tinbergen 1932). In Sweden, nutcrackers (*Nucifraga caryocatactes*) gather nuts in fall and store them in the spruce forest. The nutcrackers can find these storage sites in winter, despite snows as deep as one and a half feet, presumably from surrounding landmarks (Shetteworth 1983).

The work of Tolman and others shows us that many animals can learn more than responses and stimulus-response connections; rat, bird, and wasp learn an internal representation of the external world, a map to guide them to food or home. Like us, they recreate the outside world within themselves. These maps may be taken to imply that mind and experience are also there, at least in those lifeforms with brains that resemble our own.

Language

There is a great spiritual chasm between those people who regard mankind as superior and unique and those who think that we are merely one of many animals. Those who believe that they sit beneath the crown of nature have often cited a list of distinctive human traits that elevate us above our animal brethren—man is the tool-user, the toolmaker, the builder, or the transmitter of cultural traditions. But language is the most highly regarded human trait. Man is the speaker, the listener, the reader, the symbol-user.

Unfortunately, the list of distinctive human acts has shortened considerably in the last twenty years. Several other animals use tools—such as the sea otter (*Enhydra lutris*) off the coast of the Monterey peninsula in California. This acrobatic sea mammal brings up stones from the sea bottom, which it uses to crack mussel, clam, and abalone shells. Chimpanzees and satin bowerbirds (*Ptilonorhynchus violaceous*) actually make tools. The chimpanzee may carefully strip leaves from stems or twigs to use in termite "fishing." The chimpanzees probe the termite heaps with the twigs and eat the termites that bite the intruding twigs. Male bowerbirds, true to their name, make "bowers" to which females are drawn for courtship (Gilliard 1963; Marshall 1954). In addition to decorating its bower with blue trinkets, the satin bowerbird makes a wad of chewed bark and paints the inside walls with a charcoal and saliva mixture. Not just human structures are decorator-designed.

Many animals are builders (Collias and Collias 1976). Social insects are noted for erecting complex structures. Bees make hives and honeycombs, and certain African termites make the man-high heaps that are raided by chimpanzees. Most birds build nests, although these may vary from the mere scrapes of plovers to the elegant weavings of orioles.

Culture does not distinguish humans, either. Even flocks of birds, such as tits in England, have learned new feeding patterns that they pass on to others as a cultural trait. In the 1930s these birds learned to peck through the metal foil that sealed the milk bottles left at doorways early in the morning. This talent spread and became a permanent feature of tit behavior. Cultural traits are especially prominent in primates. The food of one group of baboons may vary from another's, even though they live in the same area.

Still, it seemed for years that language was the indisputable mark of our kind. Attempts to train apes to speak failed, even when Keith and Catherine Hayes raised the chimpanzee Vicki like a human child. Vicki could manage only four very poorly articulated "words": "mama," "papa," "cup," and "up." Subsequently it was learned that the great apes cannot control their vocal cords like a human, and therefore they will never recite Hamlet or sing rock-'n'-roll; they simply cannot articulate the sounds.

At the University of Nevada at Reno, Allan and Beatrice Gardner gambled that there might be another way to show language in an ape. The Gardners instructed the infant chimpanzee Washoe in American Sign Language. Washoe lived in a homelike environment in a trailer and proved to be an apt pupil. In three years, Washoe acquired a vocabulary of 120 signs, such as those meaning "hurry" or "gimme tickle."

Inspired by Washoe's early success, other researchers taught apes sign language. Gorillas and orangutans, as well as many more chimpanzees, have been taught to sign. Most of the teachers have been humans; however, trained chimpanzees teach each other as well. Perhaps the most accomplished of the great apes is the gorilla Koko, who has mastered five hundred or more signs. "Koko fine gorilla," as she has noted.

It should be said that there is still a hard-core resistance to the concept that the apes can possess language. Much of this resistance is based on the greater flexibility shown by signers. There is some basis for this complaint; signing is looser grammatically than spoken language, and ape signing is freer than human signing. Also, the vocabularies of the apes are far smaller than those of speaking or signing humans. Yet all the principal features are there in the signing of the apes. There is symbolism and structure; many of the signs are arbitrary and are given in characteristic order. There is even invention. When she saw her first ducks, Washoe spontaneously signed "water birds" and used her name for them thereafter.

Our position as sole symbol-user has one late challenger. In the last five years there have been two reports of accurate but limited "speech" in the African gray parrot *Psittacus erithacus.* Although several families of birds are mimics, their imitations appear to be meaningless copies. However, the gray parrot studied by Pepperberg (1983) names objects when they are presented to it.

As far as I know, the only way that we are unique is in our ability to start and control fire. We are the fire-makers; there are no others.

MAN AND ANIMALS

Scala naturae

We can easily fall into a trap comparing different species as we have been doing. We can start to think that some animals are smarter than others, that some animals are inferior and some superior. This way of thinking about animals goes back to Aristotle's conception of the *scala naturae.* According to Aristotle's notion, all animals can be placed along a continuum which ranges from the lowest to the highest. You can guess who the highest is. In the modern version of the *scala naturae,* evolution progresses from inferior primitive animals to superior advanced species. Finally, evolution reaches its end in ourselves.

The problem with this conception of inferiority in other animals and superiority in man is that it has no basis in biology or logic. All species are evolving and have evolved at least since the first appearance of many modern lifeforms in Precambrian times, some 700 million years ago. We humans have been evolving in one direction, the insects in another, the viruses in a third. We are not superior to the insects or the viruses in either the length of our evolution or in the success of our adaptations. In numbers and varieties, the insects outdistance the vertebrates. The minute virus can make the mighty fall. If the world faces danger, it is from our own species; if there is a species on the verge of extinction, it is we. We are the threatened species that threatens itself just as it threatens other lifeforms. Can we regard this as the pinnacle of evolutionary progress?

The Tree of Life

We don't have to place humanity on the throne of the animal kingdom. There is a less lordly view that is far truer to nature. Each species on earth is a point of growth on the tree of life. Each kind of us is a stem tip on this tree that stretches far back into time. On this great tree, related species emerge from common branches. From fossil evidence we can trace these branches backward to main trunks. But no trunk or branch or twig is superior to any other. Each living species, each leaf on this world-tree, is an equal distance from the great evolutionary source (or sources). All living groups are evolving outward from this primeval source, all in their own directions, all seeking evolutionary success in their own ways.

THE EVIDENCE FOR CONSCIOUSNESS

We have considered signs of consciousness in behavior: sensitivity, orientation, selection, learning, and language. Since these signs can represent conscious powers in ourselves, the signs may be taken to suggest intelligence in other species. The signs also form a scale of relationship to us. The more signs of consciousness an animal shows, the more it is like us.

Behavior is one of the ways consciousness can reveal itself. But to make the most reasonable inference about consciousness in another lifeform, we must consider all the possible expressions of consciousness: those from brain structure, chemistry, and electrical activity, as well as from behavior.

In the great external world there are animals that resemble us in all these ways, in brain structure and process and in behavior. Birds and other mammals are so like us in body and behavior that we can easily believe that these animals are conscious. A skeptic who denies the parallels between us must now bear a burden of disproof; on what reasonable grounds could he or she assert that our close animal relatives are not conscious? Just as one cannot claim to have proof of what goes on in the mind of another, one cannot deny the existence of consciousness in another being. The real question is what is the probable truth. The evidence implies that birds and mammals are conscious.

In this world around us, we also find creatures more remotely resembling ourselves. We can still see definite parallels to ourselves in them. For example, the brains of fish, amphibians, and reptiles follow the same basic pattern as our own. But there are differences in the size of brain structures. The cortex is far less prominent in these animals. Although the chemistry of their nerve cells and hormones is like ours, there are differences in the electric signals we detect in their brains. Thus in these other vertebrates, in fish, amphibians, and reptiles, we see signs of consciousness. But we also find signs of difference that are hard to interpret, given our ignorance about the site and physical nature of consciousness.

More distant from us still are the insects, the mollusks, the many forms of worms, the jellyfish and their allies, all the ancient creatures without backbones. We still see in them elements of the same physical realities that ultimately underlie or express our own consciousness. We see receptors and nerve cells, hormones and neurotransmitters, acts of sensitivity and orientations, and, in a few forms, even elaborate memories. Yet here I think we might start to feel uncomfortable in confidently asserting that these forms are conscious. Perhaps if we knew more about these animals and ourselves, we might take a stronger position. On the other hand, how could we deny these animals sentience? We have no basis to deny sentience; we can only affirm consciousness.

The earth is rich with sentience. Lord preserve it from the fires of the humans.

> *[T]here is no such thing as absolute reality. Each animal species perceives a different world from any other species. . . . The "real" world of each creature is relentlessly limited and circumscribed by the nature of that animal's information-processing systems.*
> —JOHN ALCOCK,
>   Animal Behavior: An Evolutionary Approach

> *"Whose Real World?"*
> —Title of an essay by
>   VINCENT DETHIER

8

THE NATURE OF REALITY

If the earth is peopled with other intelligences, how can we know them? If reality is more than we see on the face of the physical world, then how can we know what consciousness might lie beyond? In our culture and in the culture of our science, we have long regarded the world as made up of objects. We have been objective, indeed, but not truthful. We have been objective in our search for truth. Unfortunately, truth may not be objective. Truth is what we are and what other animals are. What we are is subjective.

How do we seek these subjective realities? How can we go beyond visible reality to approach the invisible consciousnesses of others? The only route to the invisible is through the visible. To seek consciousness we must relate it to matter. We can never forget matter; what we must remember is to go beyond it.

Matter is our only escape from the solitude of existence. Matter enters consciousness, and consciousness expresses itself in matter. Matter is our only means of communication with other lifeforms, with other intelligences.

We have seen some of the ways consciousness is expressed in matter. Our learning, our memories, our intentions can be expressed in the matter of our bodies; our behavior can thus be a copy of consciousness. Yet consciousness can also be a copy of matter. Although only certain aspects of matter are copied in consciousness, these are often just those aspects vital to our survival and to our reproduction. We smell the butter and garlic sauce for the fettucini, taste the dry red wine, and gaze into each other's eyes in the candlelight. These are important material things for us. In contrast, we are conscious neither of the speed of Halley's Comet nor of the friction in the solar wind it must pass through. These material things are remote from us and are not vital to our survival.

In this chapter we will look at how matter enters consciousness. What aspects of matter do we sense? How does material reality appear in consciousness? In the next chapter we will examine how consciousness is communicated, how consciousness reaches out to consciousness.

THE GATES OF CONSCIOUSNESS

Each of us is forever alone. No one else can ever feel our pain; we will never experience another's joy. We are exiles in our own bodies. Yet we look out on the world of matter around ourselves. We look out through our senses. Our sensory systems are the gates to the consciousness within us. Through these gates passes whatever richness or poverty we find in the great external world. Yet these gates are not broad; our senses are limited in what they detect. We sense only a few aspects of the vast material universe.

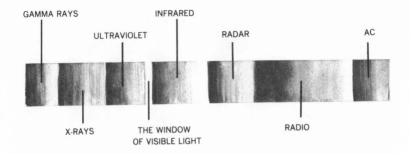

GAMMA RAYS ULTRAVIOLET INFRARED RADAR AC

X-RAYS THE WINDOW OF VISIBLE LIGHT RADIO

Visible light. Visible light represents only a small span of the great spectrum of electromagnetic energy.

What are the physical boundaries of our experience? What dimensions of the material world can we sense? We humans sense things both outside and within our bodies. We see colors, which are really a narrow band of the electromagnetic spectrum. We hear sounds, which are vibrations in the air across a limited range of frequencies—from say twenty cycles to 20,000 cycles per second. We smell an array of chemicals, most of which are volatile substances that are usually soluble in water and fat. We feel heat and cold, which reflect the kinetic energy of matter both inside and outside our bodies. We sense touch, which is related to physical tension on our skin. We have a number of internal receptors that monitor the location and state of our muscles, our oxygen and sugar levels, and many other bodily states.

The consciousness of each animal that exists has its own sensory gates, its own physical boundaries. Some animals, like sheep, share our basic set of senses. Yet there are still many differences between us: the sheep's eyes are set more widely; the sheep can see the wolf creeping up from far off to the side. Our eyes face forward and our field of vision is narrower, yet this allows us a superior sense of depth; we should be better able to judge how far the wolf is from us.

In many cases, animals have a broader range of sensitivity than we do. Cats hear the ultrasonic squeaks of infant mice. Bees use ultraviolet light to choose among the flowers.

In some cases, animals have elaborated a sensitivity to

Ultraviolet light in flowers. The marsh marigold is a plain yellow wildflower to us *(left)*. To a bee's eyes, there is a sharp contrast between the center and periphery of the flower *(right)*. Nectar guides, the lines that "point" to the center of the flower, also become prominent.

sound or light into a sensory system that is unlike our own. For example, bats, cave-dwelling oilbirds (*Steatornis caripensis*), and cetaceans use echoes to navigate and hunt. Each of these kinds of animals repeatedly makes sounds and then analyzes the returning echoes. This specialized sense is called *echolocation*. We humans do not make loud ultrasonic clicks or pings to echo off the walls in order to find our way; however, the blind can use the ordinary echoes from a tapping cane to guide them.

Echolocation can be extraordinarily precise. Bats will veer away from wire the diameter of a human hair. Dolphins have been shown to distinguish a half-inch-long gelatin capsule from a similar-sized fish at a distance of twenty feet.

The underwater echolocation of cetaceans has some special features:

> If a human diver jumps into the water with a dolphin, the dolphin can "see" inside the diver into the air passages of his lungs and respiratory system. This is because sonar sight penetrates materials that are approximately the same density as the water—like human flesh—and returns different echoes from objects with different densities. The greater the difference in density, the more easily sonar can discriminate. In the case of the diver, his lungs show a greater contrast to the water than his wet suit. To the dolphin, the diver might look like an x-ray photograph of the human body [Warshall 1974, 134].

John Sutphen (1974) has suggested that cetaceans can thus sense the internal states of their fellows; a dolphin can sense whether another is ill or tense, for example.

Some of the physical dimensions of other animals' sensory worlds are entirely alien to us. Their senses detect physical qualities that we are not aware of in our normal existence.

The Electric Fish

Consider this scene. It is a warm night on the edge of the Brazilian jungle. A man walks slowly along the edge of a stream. He stops and puts two slender steel electrodes into the water. The electrodes are connected to a small amplifier the size of a pack of cigarettes. The man listens over headphones to the amplifier as if it were a Walkman. The sound he hears is not that of music; instead, he is listening to the electric signals broadcast by different kinds of knifefish. The South American knifefish are weakly electric fish, fish that broadcast a low-voltage signal into the water around themselves. Other weakly electric fishes include the freshwater African mormyrids and the common saltwater skates.

The knifefishes and the mormyrids usually send out a continuous series of pulses (C. D. Hopkins 1974, 1981). Each pulse consists of a pattern of current flow between head and tail. Any living fish or insect distorts the flow of current, since it has a resistance lower than that of fresh water. Rocks or other objects with a resistance higher than that of the water also distort the field of flow. Thus the knifefishes can use the pulses of current to find prey or to navigate in their dark and often turbid environment.

Skates discharge infrequently, and so one might suspect that their signals convey some social message. In line with this, individual skates will discharge more frequently when held in groups. The signals of the knifefishes and mormyrids can also serve a social purpose. For example, in Guyana, a male *Sternopygus macrurus* knifefish will "hum" along at a low discharge rate in his hiding places under banks or rocks. However, when

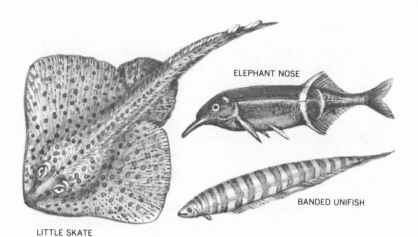

ELEPHANT NOSE

BANDED UNIFISH

LITTLE SKATE

WEAKLY ELECTRIC FISHES

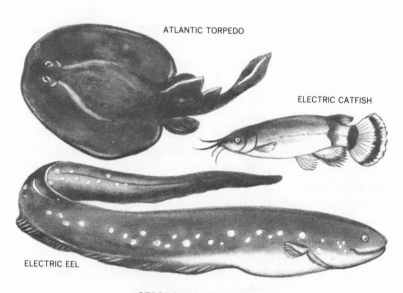

ATLANTIC TORPEDO

ELECTRIC CATFISH

ELECTRIC EEL

STRONGLY ELECTRIC FISHES

Electric fishes.

one of the more highly pitched females swims by, he "wolf whistles," as it were, by dramatically changing the frequency of his own signals.

Although the electrical nature of the weakly electric fishes was only recently discovered, the discharges of the strongly electric fishes have been known for thousands of years. In ancient Rome, the physician Scribonius Largus recommended that the sufferer from gout place his swollen foot on a live black torpedo, a species of electric ray. Presumably the patient's leg became numb as the torpedo's forty-five-volt discharge surged through the foot. Since a similar treatment with the electric eel *Electrophorus electricus* was used by South American Indians (Wu 1984), it may well have been an effective cure for gout; in any case, it certainly demonstrates the power of strongly electric fishes. For drooping eyelids the medieval Chinese applied shocks from an electric catfish, *Parasilurus asota*, which directly stimulated the affected muscles.

Naturalistic observations show that all three kinds of powerful electric fishes use their discharges to stun prey. As Aristotle, the father of naturalists, noted:

> The torpedo narcotizes the creatures it wants to catch, overpowering them by the power of shock that is resident in its body, and feeds upon them; it also hides in the sand and mud, and catches all the creatures that swim in its way and come under its narcotizing influence.

In addition, their shocks can serve as defenses. The six hundred-volt shock of the electric eel, for example, can topple a horse fording a stream.

Both kinds of electric fishes have special sensory cells that detect electric signals, as might be expected. The electrosensory cells also permit some of the fishes to detect the weak electric fields generated in water by the bodies of all living creatures, including yours and mine. Certain nonelectric fishes such as sharks also have similar electroreceptors. Dogfish sharks, for example, can locate flounders buried under sand by sensing their minute electric output.

Electroreception can be quite elaborate. For example, the elephant nose (*Gnathonemus petersii*) has three separate sets of electroreceptors on the surface of its skin. One kind of receptor picks up the weak direct-current fields that any living body generates in water. A second kind detects any distortions that may be present in the field set up by the electric fish's own discharge. A third turns off during the fish's own discharge but does respond when any other electric fish discharges. In other words, the elephant nose has separate receptors for faint DC and AC signals, as well for testing resistance in the water around itself.

To summarize, the physical world of the electric fishes is radically different from our own in two ways. First, they can sense low currents of electricity. Second, they can produce electricity, in the form of signals or shocks. Neither of these powers is given to our bodies. It has only been in the last two centuries that humans have invented instruments to do these same things. We have devised voltameters, ammeters, and oscilloscopes to detect and describe electrical outputs. We have designed a variety of generators to create electricity. It is because we have invented these things that we can understand the physical worlds of the electric fishes and also get some inkling of what their consciousness might be like. In a sense, our inventions are like the analogies we can use to understand other creatures.

THE STREAM OF CONSCIOUSNESS

We have seen that each consciousness has a set of physical boundaries. Through the senses, each consciousness can explore limited aspects of the material world. Matter thus reaches mind, and mind reflects on matter.

Yet what form does matter take in mind? How does the physical world appear in consciousness? How does the dark

Amazon seem to the knifefish? How does the winter sky feel to the migrating songbird?

This is a great mystery. How does another animal experience the world? What is physical reality to another lifeform? Indeed, what is reality for another human? We have no way of knowing directly. The only answer for us lies in observations of the material world.

We know that even the smallest organisms mirror the outer world. Consider the external world of the bacteria:

> A bacterium's habitat is so foreign to our own intuitions that it is necessary to spell out what it feels like to be an organism that is only two or three micrometers long and one micrometer wide. Being so small, [its] inertia is insignificant and the only forces a bacterium experiences are those of viscosity. Scaled up to our size, it is as though [the bacterium] had to move through liquid asphalt [Collett 1983, 41].

This liquid asphalt is formed of molecules, and these battering molecules are what the bacteria must contend with. Bacteria have evolved to find and stay near concentrations of favored food molecules. A familiar example is the ubiquitous bacterium *Escherichia coli* that resides in our guts and in all the polluted waters of earth. On the surface of *E. coli* are thirty kinds of receptors, each sensitive to particular chemicals. *E. coli* can even sense mirror-image forms of several sugars. Such molecules can tell *E. coli* to change direction as it swims. The fewer the changes, the more the bacterium moves in one direction, which is a roundabout way to find food. Thus the cell surface of *E. coli* has receptors that reflect the chemistry of this bacterium's world; moreover, its behavior itself reflects the distribution of food, as well as other chemicals.

In addition to their sensitivity to chemicals, many kinds of bacteria are sensitive to light. It was recently discovered that certain aquatic bacteria are even sensitive to magnetic fields (Blakemore and Frankel 1981). Within their bodies are small crystals of magnetite that cause them to orient along the lines of magnetic forces generated by the earth. Magnetite is found in the lodestone used in the first human compasses. Although humans have had compasses since the Middle Ages, it was Sir William Gilbert, the physician to Queen Elizabeth I, who first proposed that the earth acted as a magnet. The bacteria had "known" all this for hundreds of millions of years before Sir William. But why should bacteria act like tiny compasses? A compass can not only tell north, but it can usually tell up and down, since the lines of magnetic force dip as one approaches the poles. The bacteria can usually follow their compass orientation to swim downward to the food-bearing muck below.

It is remarkable how the outer world is reflected in the bacteria. Yet the internal representation of the outer world is even more elaborate in larger animals. There are two ways to infer what the inner representation is like: through the brain and through behavior.

The Mouse in the Brain of the Rattlesnake

In the brains of rattlesnakes and pythons there are pictures of the world. One kind of picture is generated by the pit organs of these snakes. For a long time no one understood why these snakes had prominent pits on their snouts. Then, in the 1930s, it was discovered that rattlesnakes will strike at warm lightbulbs covered with black cloth. They will do so even if their eyes are covered, but they will not strike if their pit organs are covered. How do these pits let the snakes sense the invisible infrared light or heat that comes from warm objects?

The pit organ acts as an infrared pinhole camera, in that the infrared radiation forms an image in the pit like the image formed in the eye, though hardly as sharp. The impulses from the pits pass through a series of stages in the brain. At each stage the infrared image of the world is re-created in a map. At the back of the midbrain, this map of infrared reality is drawn right on the surface of the brain. The front of the infrared map is at the front of this surface area, the sides on the sides and the back at the back, just like a miniature painting. This infrared map overlaps a second map, one that comes from the eyes to cover

Pit "vision" in a sidewinder. This rattlesnake
detects warm or cold objects that move through
its two conelike fields of infrared sensitivity.

the same surface (Newman and Hartline 1982). Individual cells
in this surface respond to stimulation of the eyes or pits alone,
or to combinations of this stimulation. One kind of combination
cell responds to moving, warm, illuminated objects—a mouse,
for instance. Another kind of cell fires in response to cool, moving stimuli—such as a turtle waddling in the grass, but fires less
if the object is warm.

The image of the rattlesnake's consciousness is conjecture.
We know that the snake acts on this information, but we don't
know that the snake's experience consists of images. This is,
however, a common way to conceptualize the consciousness of
animals. We know that people both have images and think in
words; since animals don't talk, we may suppose that leaves
them with images. This is certainly a possibility; perhaps most
animal consciousnesses are cast in images. There are other possibilities, however. A snake may "think" in feelings or in inten-

Mouse in the mind of a sidewinder. At the bottom is shown a view seen by the snake's eyes. The center represents the same view "seen" by the infrared sensitive pits. The top illustration suggests the accentuation of moving things in the visual field that might be made by the combination cells.

tions. Nonetheless, there is some evidence to support the hunch that at least some animals think in images. Let us consider an example.

Stars and the Behavior of the Bunting

In Michigan, the alder leaves begin to turn yellow in September and October. Small songbirds, like the indigo bunting (*Passerina cyanea*), begin to put on fat for the long journey to their wintering grounds in the Bahamas, Mexico, and Central America. Many of these songbirds will travel by night, some flying over bodies of water as vast as the Gulf of Mexico. How do such sparrows and warblers find their way?

Years ago it was noted that captive songbirds became intensely restless as the time for migration neared. The young zoologist Stephen Emlen (1969, 1975) used this restlessness to help find out how night fliers navigate. Emlen captured indigo buntings and held them at night in odd conical cages. The tops of the cages were open to the night sky; the sides were lined with paper; at the bottom were ink pads. As the time for migration came, the buntings began to hop up on the paper-covered walls of these cages. In the morning their footprints left a trace of all their jumping. On clear nights, the hopping was clearly oriented southward. On overcast nights the hopping went in all directions. The buntings were orienting by the night sky. What about the sky was giving the buntings their migratory heading?

To experiment further, Emlen took his buntings into a planetarium where he could control the sky, as it were. There Emlen showed that the buntings were orienting by the constellations around the North Star. No particular constellation was critical: if the Big Dipper was missing, the Little Dipper, Draco, Cepheus, and Cassiopeia would do. It would seem that the buntings were flying by the very image of the stars. But perhaps this was not an image at all. It might be instead that this knowledge of the stars was unconscious, a primitive and mechanical connection of points of light with migratory flight movements handed down through some mindless process of evolution.

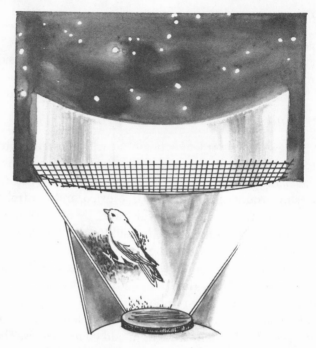

Buntings under a false sky. Indigo buntings can learn
the constellations in a planetarium. Hopping up and
down from an ink pad, they leave a record of their
migratory heading on the conical paper sides of their
special cages.

We can ask if buntings are in fact born with a knowledge
of the heavens. Apparently this is not the case. With his star
projector, Emlen showed that the buntings learn the heavens
in their first summer. Buntings exposed to a "normal" planetarium sky oriented correctly southward; those reared under a
blank sky showed no pattern of responses. In fact, the buntings
could be trained to fly to Betelgeuse, a star in Orion, if that was
projected as a false Pole Star. Why should the buntings learn the
stars that wheel around the Pole? The heavens are not fixed; the
earth wobbles over time like a top. In thirteen thousand years
the Pole Star will be Vega; Polaris will then lie 43 degrees north.
A rigidly fixed navigation system would fail under these shifting

stars; learning lets the bunting have the flexibility always to find true north.

The detail of the bunting's map and their learning of the skies suggest that they may indeed be conscious of an image of the stars. Yet we cannot be certain. We cannot enter the consciousness of a bunting or of a rattlesnake, or of any other being. But it is clear that buntings and rattlesnakes act as if they had images, images of the stars or of heat. Perhaps when we know more about the brain we will be more sure. Yet it would not be surprising if animals were conscious of images. That is how much of the external world appears to us. What better way to recreate the world around us than its own image? What better way to represent the geometry of our material world?

THE KALEIDOSCOPE OF REALITY

Beyond the face of nature is concealed intelligence—our own and that of many other lifeforms. We can ignore this hidden universe for the certainty of the familiar world of things; or we can venture toward the great subjective world of reality beyond, which is our own reality. Perhaps we should be brave and seek this subjective world. It is what you and I are. It is what many other lifeforms must be. Familiar but unfamiliar, knowable but never knowable, visible but invisible.

Each intelligence that we encounter is a universe within a universe. No two sets of senses are identical; no two nervous systems are the same. Each consciousness senses and perceives different aspects of our shared material world. In this way, each lifeform is its own definition of the outer universe that encompasses us all.

9

THE COMMUNICATION OF CONSCIOUSNESS

Humpback whales (*Megaptera novaeangliae*) sing on their winter breeding and calving grounds (Payne, Tyack, and Payne 1983; Baker and Herman 1985). Their long and complex song is shared by almost all the singers on the grounds. But as the season passes, the song changes. Singers take up the new or changed phrases of the song and let the old ones drop. Over the quiet summer hunting period, the song stays virtually unaltered.

This haunting winter song is a signal for other humpbacks. Somehow this communication is passed along over great distances. The season's song will ultimately be sung by all the humpbacks in a vast area such as the North Atlantic or the eastern and central Pacific (Payne and Guinee 1983).

There are two facets to whale song. First, the song expresses something about the singer. In the case of the hump-back, we know that singers are lone males and that they sing in winter (Darling, Gibson, and Silber 1983; Tyack and Whitehead 1982). What the song says about the singer is called the *message*. The message of the whale song is something like "I am a lone male humpback in breeding season."

Second, the song affects other whales. The response of other whales to the song is its meaning. With the humpback, singing stops as soon as the singer is joined by another whale. The second whale never sings (Tyack 1981). This response to song may represent a cow coming to mate with the singer. This is not certain, however, since the sea conceals the sex of the whale joining the singer, and in any event, no one has ever seen a humpback mate. We cannot be sure of the precise meaning of humpback song, since we have not seen the exact responses that song leads to; the sea thus keeps the meaning of humpback song mysterious.

All communication follows the basic pattern we see in the whales (W. J. Smith 1965, 1977). There is a sender, a signal, and a receiver. The signal expresses something about the state of the sender; this is its message. The receiver interprets the signal in its own way; this is the meaning of the signal.

Let us consider a less remote example of communication: the song of the yellowthroat (*Geothlypis trichas*) pictured on the following page. The thickets and tangled brush along swamps, streams, and roads in the eastern United States and Canada are quiet in winter. In spring, this changes as the yellowthroat returns to its brushy northern home from the West Indies. The black-masked male starts to whistle "witchery, witchery, witchery, witch" from low perches in the thick growth. The males establish little territories and challenge other male yellowthroats that try to enter. The territory-holder is soon joined by a drabber female (or two). The pair mates and builds a nest, the eggs are laid and hatched; the song of the male becomes less frequent after spring lapses into summer.

We understand the songs of birds better than the songs of whales. Birds are easier for us to observe; their songs are easier

Yellowthroat.

to hear. We know a great deal about the internal states of these singers. We know they are typically males and have a special center in the brain devoted to singing. We know that in most species testosterone has begun to flow in their veins in the lengthening days of late winter. Along with the flow of testosterone there are many changes in behavior. When we hear the repeated song of a yellowthroat along a country road, we know that there is a male in his brushy territory, that he will attack rivals and will attract females; he may even scold us, yellowthroats being bold for their fraction-of-an-ounce size. Thus when we hear yellowthroat song we know a lot about its message.

We can also say the song of the yellowthroat has meaning for us, even though we're not birds. There are many common human responses to the song of the yellowthroat; for example, we may hear the three or four phrases of the song and recognize that this sprightly three-quarter-time tune is a sure sign of spring. In contrast, if we were birds, the song's meaning might be quite different. If we were another male yellowthroat, the meaning of the song might be aggression; if we were a female, the song might mean attraction; if we were a sharp-shinned hawk, the song might mean food. But this is just speculation. It is not simple to say what the meaning of the song is for these other animals. We know more about the message of yellowthroat song than we do the meaning for other birds.

To understand the meaning of song for other birds, we must make careful observations of the receivers of these signals. We must watch rivals, mates, young, or predators. We must try to specify and understand their response to the song. For example, if we were to play the tape-recorded song of a yellowthroat along a brushy swamp, any resident male would rush out to attack our simulated intruder. Playing birdsongs like this is an excellent way to see many secretive species up close. The sound of a rival makes soras and rails and other timid birds bold; they will walk all around the tape recorder, apparently searching for the intruder. There is even one case in which imitated calls seemingly established a territory. I once knew a dean of forestry whose hobby was "calling" owls. Along one favorite woodland road, he repeatedly hooted up a horned owl. After a few weeks the owl would no longer come out to the road, but would still come if the dean hooted farther back in the woods. The dean had inadvertently set up his own owl territory along the edge of the road. These observations and many others show us that to a territory-holder the meaning of bird song is "invader," and that "invader" means "search and attack."

Bird song ranges from simple calls to elaborate repertoires. We humans often find the sounds beautiful—that's another part of the meaning of bird song for us. We wonder why some singers have such a range of songs. What is the meaning of a variety of songs for rivals or mates? In England, the zoologist John R. Krebs and his associates studied this question in the great tit (*Parus major*), a close relative of the North American chickadees (see S. J. Smith 1985). The message of the varied songs of the great tit can certainly be subtle. In Marley Wood near Oxford there is a population of great tits that has been carefully watched for years. Some of these great tits had only one or two songs; others had three or more. Peter McGregor, John Krebs, and Christopher Perrin (1981) found that singers with a larger number of songs left more breeding offspring. The singers with an average or greater than average number of songs also fledged fatter offspring. Only careful, patient study revealed

Great tit. (After Joan Parsons.)

that evolution was favoring the singers with more songs. But why?

Krebs, Ruter Ashcroft, and Mike Webber (1978) removed male great tits from their territories in Higgins Copse near Oxford. As expected, undefended territories were quickly occupied by other males; however, if a tape recording of great tit song was played, the invaders delayed their entry. The greater the variety of songs on the tape recording, the slower the invaders. One part of the meaning of the great tit songs for invaders is therefore "keep out," and a greater variety of songs makes this prohibition stronger. One reason for the variety of songs, then, was the meaning many songs had for rival great tits.

While the meaning of bird song may be "keep out" to an intruder, what is the meaning to a mate? While we know that many singers find mates, there is unfortunately simply no evidence that unattached females are drawn to song by itself. However, two other meanings of song for mates have been noted in captive birds. Female song sparrows (*Melospiza melodia*) invite males to mount with a distinctive courtship solicitation display. Caged females ready to mate give this "come hither" display even when they hear recorded songs. The greater the variety of songs played, the more likely the courtship solicitation (Searcy and Marler 1981). Captive female canaries will build nests faster and larger if they hear a greater variety of song (Kroodsma 1976).

This returns us to the humpback. Perhaps the most unusual feature of humpback song is its changing nature. Throughout the breeding season the collective song changes. With the exception of our own species (and possibly of the yellow-rumped cacique [*Cacicus cela*], a relative of the oriole), all other songs in nature remain relatively fixed once they have developed in an individual. Why should the humpback song change? Perhaps the humpback cows are more attracted to new songs than old, just as female song sparrows and canaries are more responsive to more kinds of songs. This preference for variation may be a link between human, bird, and whale. Perhaps there is some affinity in our nervous systems that lets variety in song have meaning for us all.

Conscious and Unconscious Communication

Until now we haven't asked whether messages and meanings may also reveal consciousness. It is certainly true that the displays of animals have developed or evolved for purposes of communication. It is also clear, however, that communication can take place on either a conscious or an unconscious level. The sender may or may not be conscious of the message, and the receiver may or may not be conscious of the meaning of the display. Consider our own communication. Often we consciously try to communicate our feelings or thoughts to others. At the same time, we also quite unconsciously communicate things about ourselves through our posture, expression, gaze, or distance from others. Our unconscious communication can reveal either conscious or unconscious aspects of ourselves. It can expose aspects of our consciousness that we don't want others to know; it can also tell others things about ourselves that we are unaware of.

Let's elaborate on the possible roles of consciousness in communication by putting ourselves in the picture as observers. Imagine that we are watching California sea lions (*Zalo-*

phus californianus) on a small offshore island. The bulls seem to be barking incessantly. With the aid of a telescope we can see that the bulls move about and chase or intercept other bulls that approach their stretch of beach. The sea lion females, half the size of the males, seem to ignore both the barking and the beach boundaries enforced by the bulls on one another. What is the message of the barks, and what is its meaning? Is either the message or the meaning conscious?

In the table, which follows, the possibilities for conscious communication are outlined. The territory-holding bull may or may not be conscious when he barks. If he were, his conscious state could be part of the message of the barking. The bark might reflect a state like anger, for example; one message of the bark would then be anger. If the bull were unconscious, however, the barking would denote some unconscious state. (It is possible that a radical behaviorist might insist that a roving bull was not conscious, although I can't imagine even a behaviorist asserting this. More likely the behaviorist would say that its mind didn't matter, which could be taken as a play on words.)

Whether conscious or unconscious, the internal state of the bull is *encoded* in his barks. In other words, the bull's internal state is expressed in certain external ways. This encoding may be conscious or unconscious. That is, the bull might be aware of anger but may or may not be aware of the acts that go along with his anger. This is not altogether farfetched. When we humans are angry, we may be unaware of our flushed faces or tensed muscles; these can be unconscious signs of our anger. On the other hand, we are often aware of our words and the tone of our voices. We are conscious that these signals encode our anger. Perhaps the bull is, too.

The bark is a signal that may be heard by other sea lions and by ourselves. If the bark is heard, then the signal is *decoded*. Hearing may be on a conscious or an unconscious level. For example, if we humans were walking along the shore, we might be aware that something was odd. Only later might we realize that we had been just barely hearing the doglike calls of the sea lions over the sounds of the surf. Many communications are

COMMUNICATION BETWEEN HUMAN AND ANIMAL

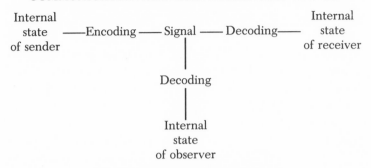

Any internal state or coding process
can be either conscious or unconscious.

decoded unconsciously by humans; we often call this intuition.
We have an intuition about another person; we know he or she
is angry, but we can't say why. Perhaps sea lions decode barks
the same way; they react but are not aware why, although there
is no reason for us to think that the sea lions would not consciously hear the barks.

Our reaction to the barks is the signal's meaning. For another bull on the island, this reaction may be either conscious
or unconscious. The other bull might react with fear, for example. But this fear might not be conscious. This may sound con-
tradictory, but we humans can certainly be afraid and only
realize our fear after the fact. In any case, we can't know directly whether the intruding bull is consciously or unconsciously fearful.

The diagram of consciousness in communication is a more
elaborate version of the simple description of communication
given at the beginning of this chapter. All of the elements in the
diagram can represent conscious or unconscious processes, except for the display itself, which is an observable event. How
can we discover what aspects of communication are conscious?
The answer is by analogy, and there are several approaches that
may be helpful to us.

The Nature of the Communication

We cannot say directly what aspects of communication in other animals are conscious. Yet there are several kinds of evidence that bear on this question. One is the type of communication. In certain contexts, we humans are quite conscious of events. When our life is threatened, for example, we become exquisitely aware. We tend to be very attentive when we try to attract a mate or fend off some rival; we call the associated internal states love or jealousy. Most parents are very conscious of where their children are; if we can't find our children, incredible anxiety can result. Intensity, love, jealousy, fear—these internal states can rule our consciousness. These strong feelings can also dominate our communication.

We know that the songs of yellowthroats and whales are also linked with mating, and this is one of the times when many animals communicate with one another. Mating means that animals must come together, and signals can make this happen. The birth of young may necessitate further coordination between parents as well as between parent and offspring. Both parents and offspring are likely to communicate. Each ewe and each lamb has its own distinctive "baa"; this call summons the ewe to the lamb and the lamb to the ewe.

The presence of danger is also a time when signals are common. Many animals give warning when predators are detected. Many species of deer erect their tails when they detect the movement or the scent of a predator. Each species has a distinctive pattern of rump and tail coloration that emphasizes this signal (Guthrie 1971). Animals may also have signals to confuse or threaten predators. In India the gaur, the largest member of the cow family, threatens the tiger by snorting and shaking its massive head and horns.

In all of these contexts, the survival of the individual and ultimately of the species is at stake. Communication is common in life-or-death situations. If there were ever times when animals were sentient, these situations of danger and care-giving would seem the most likely ones. It would be here that alertness and feeling were at their extreme. Of course, this is a simple argument based upon human experience.

Evolution of Communication

Another way to link consciousness and communication is through evolution. Many displays have been formed in evolution in the process called *ritualization*. It is as if evolution took some act of an animal and then exaggerated and standardized it so that it became a signal. Consider the courtship solicitation display of many songbirds, such as the song sparrow. The female fans her wings and raises her head. This display comes from the begging behavior of nestlings, and somehow got transposed into courtship.

One of the most ancient and widespread displays is the *vocalized bared-teeth* threat. Many animals threaten their own kind or others by opening their mouths, baring their teeth, and making a sound (van Hooff 1972). A familiar example is the snarling, menacing watchdog. A less common example is the gape of the opossum. The opossum has quite an array of teeth in its jaw; as it gapes, it holds its mouth steadily open and hisses. The opossum doesn't actually attack us. Instead, it keeps its mouth wide open in a formal, standardized gesture of threat. It is as if evolution took a single element of aggression and made it the sign of threatened attack.

The vocalized bared-teeth display is shared by our fellow primates. The display is seen when an animal is threatened or attacked by a more dominant animal. Below, a subordinate rhesus macaque female (*Macaca rhesus*) is threatened by a higher-ranking female, and responds by baring her teeth and geckering. The vocalized bared-teeth display is seen in our species, too—we call it the scream.

In some of our closest relatives, in the macaques, baboons, and great apes, there is a related *silent bared-teeth* display. In most species this display is given by a lower-ranking animal when confronted with one of higher rank. In the Barbary and Celebes apes, in some baboons and in the chimpanzee, the

Threatened female rhesus monkey. This female bares her teeth as she "screams."

silent bared-teeth display can also be given by the dominant animal, seemingly as a sign of reassurance.

What do we humans do when confronted by a higher-ranking person? When we greet the boss, we bare our teeth silently in a smile. And, as in the baboon and chimpanzee, this dominant figure may smile in return.

The bared-teeth expressions are tense; feel the tension in your facial muscles as you smile. In contrast to these tensed expressions, most primates have a special play-face. This *relaxed open-mouth* display is given during play-wrestling in many kinds of monkeys and baboons. In fact, it may have been ritualized from the gnaw-wrestling of rough-and-tumble primate play. Along with the relaxed open-mouth display goes fast, hard breathing, which is voiced in some species. In the chimpanzee, the fast breathing is often accompanied by a "soft, low-pitched, noisy 'ah, ah, ah'" (van Hooff 1967). Tickling the chimpanzee is a good way to get it to make the display. We humans have our own word for this display—we call it laughter.

Grins and smiles are human displays, traceable through a series of intermediates to the primal open-mouthed threat of the mammals. Our play-face is a display that seems to have evolved in contrast to the social and muscular tension inherent in most of the bared-teeth expressions. Along with these human displays go feelings of attachment and submission, feelings that are quite conscious. When we see other animals bare their teeth in conflicts or in play, we are reminded of ourselves. The analogy in our expressions suggests that the other animals may be conscious during their displays and perhaps that they even feel somewhat like we do. This interpretation is reinforced by the shared ancestry of such displays.

SPEAKING WITH THE ANIMALS

There is a third kind of evidence about consciousness that can be gained through communication. We can always ask the animals. Our questions can take different forms.

What Birds See in Us

Some of our questions are experiments. For example, we can teach a pigeon to peck at a stimulus to receive grain. Then we can test the pigeon to see what it has learned about the stimulus. If a pigeon is trained to peck a red light, it will also peck other colored lights, but not so often. In other words, the pigeon has a "concept" of red, which generalizes somewhat to other lights.

We can teach the pigeon more natural discriminations. We can readily train a pigeon to peck only at keys on which other pigeons appear. A pigeon can also learn to peck at keys showing water. We can even teach a pigeon to peck at keys showing individual humans. (We are not doing this to train extras for Alfred Hitchcock movies, but to show that pigeons are capable of people perception. Actually, Hitchcock's birds were trained

Piping plover.

to lick peanut butter from faces.) Pigeons in parks presumably use their recognition skills to spot potential feeders.

Pigeons are not the only birds to recognize humans. On sandy beaches along the Atlantic, the piping plover (*Charadrius melodus*) makes its nest in direct competition with human beach-goers. This pale shorebird defends its scrape of a nest by intercepting humans or their pets and leading them away from the nest, often with a broken-wing display. One remarkable thing about the piping plover's heroics is its recognition of individual humans. By their clothing the plover recognizes humans that have previously disturbed the nesting area; it even reacts to the direction in which a human's eyes are looking. Carolyn Ristau and Donald Griffin argue that the piping plover's defense of its nest is so elaborate that the plover can only be conscious (Jolly 1985). They feel that the complexity of the defense demands conscious powers of learning and thinking. To imagine an unconscious automaton that decoys predators as the plover does is preposterous. It's easier to think, or so the analogy to ourselves suggests.

Signing with the Apes

There is another way we can speak to the animals. With the great apes, we can ask questions about consciousness directly. As we have seen, chimpanzees, orangutans, and gorillas can be taught symbolic languages. Great apes have learned to read sequences of plastic tokens, to punch banks of computer controls, and to sign with their hands. Since the symbols are shared between us, we can "speak" to the chimpanzees, orangutans, and gorillas. This communication is not perfect, since no doubt the symbols mean somewhat different things to man and ape. Still, there is also clearly some common meaning, some similar response to the symbols, and this commonality lets us communicate back and forth with a nonhuman lifeform in a language of our invention.

Humans have had many views about the character of the apes. In our Western tradition, we have often thought of the ape as a savage and inhuman brute. But this is not a universal attitude: in West Africa there was a belief that lowland gorillas were a kind of men who chose not to speak in order to avoid taxes. This latter view is closer to the truth; the apes are rather like us and they do escape the tax collector.

American Sign Language (ASL) has been the most usual way of communicating with the great apes. ASL is not merely a version of English, but a living, evolving language with some half-million users. Its signs are not exact equivalents of words (see Waterhouse, Mindess, and Silbering 1984). The shape of the signs, as well as how and where they are made, modulate their meaning. For instance, if the sign for "give" is moved toward the signer, it means "you give to me"; if the sign is made moving away from the signer, it means "I give to you." While one sign can mean many things, the signs can also be very precise. The English word *run* has many meanings—for instance, your nose may run or you may run down the street. ASL has thirty-four different signs that stand for exact meanings of the word *run*.

Since ASL lacks articles, prefixes, suffixes, and the verb "to be," translation can be imprecise. Literal translations of sign language may sound crude when in fact it is the English that often inadequately captures ASL. Thus, when we read what the apes sign, we must remember the apes are not speaking English, but a language as remote from our own as Chinese. Actually, Chinese is in some ways more like ASL than like English.

The fact that apes use a "foreign" language has led to various technical disputes and misunderstandings by English speakers. I think we can ignore these disputes as long as we remember that sign language is not English and that ape sign language may not be exactly like ASL.

The first signing ape was the chimpanzee Washoe at the University of Nevada. Washoe later moved to a chimpanzee colony in central Washington. There she has helped teach others to sign. The young male Loulis learned to sign exclusively from other chimpanzees; for the first five years of his life, his human caretakers carefully avoided signing before him. Loulis was taught to sign quite proficiently by Washoe and the others; he may have thought his humans weren't too bright, though.

There are only two gorillas that have been taught to sign: Koko and Michael. Koko and Michael live in Woodside, California, near San Francisco. Koko and Michael have been signing for years and have always been under the guidance of Penny Patterson. Penny has moved out of university circles to create a foundation to undertake the lifelong study of communication in these two remarkable lowland gorillas. The language skills of these animals has given us our best access to nonhuman consciousness.

Let us read a few samples of gorilla signing. Note that finally the tables are turned. We have talked on and on in our words about the animals; now for the first time we must listen to what the animals have to report about themselves. We will quote the symbolic communications of apes in the language of the deaf. We can relax because what the apes tell us is childlike.

What do Penny Patterson's gorillas like? Do they like to work?

> BARB HILLER: Want to do a little work?
> *Koko brings a blanket over and sits on it by me.*
> KOKO: Me gorilla tickle you.
> *We have a tickle session.*
> BARB: Work?
> KOKO: Gorilla me knock there (pointing to the floor).

KOKO: That trouble there (pointing to a Smurf doll that I've put into the mouth of a Tyrannosaurus Rex toy).
BARB: Can you save the baby?
Koko takes the doll out of the dinosaur's mouth and kisses it.
BARB: You saved the baby.
Koko laughs and puts the doll back into the dinosaur's mouth.
(Hiller 1984b, 10)

Actually, gorillas do sometimes like to work. Consider the responses of Michael in the free-association test given at the start of this chapter. Freud and Jung would have been startled at this use of their association techniques.

Koko can shower affection on her trainer, her dolls, and on her pet cats. Koko refrains from hurting her pets, being very gentle despite her great strength and a kitten's provocations:

KOKO: Cat red bad Koko.
BARB: What happened?
KOKO: Teeth visit gorilla. (Hiller 1984a, 9)

Koko can also protest her treatment by her human trainers, and she does this, quite spontaneously, with insults. In fact, you can ask Koko about this:

BARB: What do you say when you really want to insult people?
KOKO: Dirty.
BARB: OK, can you think of another one?
KOKO: Sorry, gorilla polite.
BARB: It's OK to tell me.
KOKO: Toilet. (Hiller 1984a, 9)

Our conversations with the great apes have taught us many things. We know that apes recognize themselves and can use impressive symbolic and logical skills to get around in the world. More than these "intellectual" talents, the great apes have a life of feeling that is uncannily like our own. They can be angry, afraid, or hurt; they can threaten, hug, or complain. They can cherish, nurture, or be possessive. They can be "good" and they can be "bad." They can joke with us and laugh.

With these messages from the apes, we find we are no longer the only speakers in the universe of our symbols. And with our own symbols the apes let us know that we are not alone in the world of our feelings.

WHAT THE ANIMALS TELL US

Almost all animals speak. They speak in codes that we do not understand at first. They speak about danger and mates and young. They speak of life and death, of life coming out of life. Much of their passion is embedded in these signals that they send out to the world. With all our intelligence we hear these signals and strain to decipher the codes. We live in a world of messengers that we have only begun to hear.

The valley spirit never dies;
It is the woman, primal mother.
Her gateway is the root of
heaven and earth.
Use it; it will never fail.
—LAO TSU, Tao Te Ching

10

THE IMAGE OF THE SUN

We've been sitting here too long. Let's walk.

The path winds through the tall grasses to the woods. The ground is rough—small red and brown stones in a bed of coarse volcanic soil. The colors in the stones do not exist outside of ourselves; they are the creation of our consciousness. What is really outside ourselves is matter, and the soil and rocks we see are but configurations of mass and energy. The colors we perceive are the wavelengths of radiant energy not absorbed by the shimmering forms of matter around us.

The world of physics through which we pass is one of law, power, and illusion. We must obey its rules. The forces in a single atom are awesome, yet we see solids where there are none; we see permanence where there is only transience. The basalt rocks we are stepping over, the ashlike volcanic soil, the valley itself are really almost an empty void; yet we do not fear we will fall through the thin vapor of spinning electrons, neutrons, and protons that actually make up the valley floor.

If matter is not what it seems, what is it? Could there be some mental aspect to matter, which is lost in our commonplace and comforting perceptions of earth and stone? We cannot count the trillion trillion atoms of air in our lungs, even though they are the breath of life. We have yet to catalog the most elementary units of matter, although our thought depends on such particles. We cannot conceive of how mass and energy are transformed into consciousness in ourselves. How then can we deny that matter may also be mental, just as we are matter and mind?

> As the God of the Old Testament asked Job:
> Where were you when I laid the foundation of the earth?
>   Tell me, if you have understanding?
> Who determined its measurements—surely you know!
>   Or who stretched the line upon it?
> On what were its bases sunk,
>   or who laid its cornerstone,
> when the morning stars sang together,
>   and all the sons of God shouted for joy?
>   —JOB 38:4-7

As we walk, sunlight filters through the spreading oaks and arching bay laurels along the path. It backlights the showy white flowers on the overhead branches of the buckeyes. The more modest flowers of the toyons beneath promise bright red fruit in late fall. All of these plants live to capture the sunlight; indeed, the twists and turns of their limbs are the history of their search for light. Like us, the plants can act and sense the outer world. They also procreate, and for this purpose the buckeye and toyon lure bees to carry their pollen.

Are plants sentient? Can these lifeforms feel as they sense, reproduce, or communicate? We may not think so because we know our consciousness resides in nerves. Yet what is it about neurons that makes them the source of our consciousness? The impulses that travel down our nerves are sequences of ion flows; the slower potentials in our brain are the sums of shifting ionic balances. If our consciousness ultimately comes from the flow of ions, then we must acknowledge that ions also electrify the cells of plants. Perhaps the oaks and buckeyes do not need neurons to feel; the voltages that we can measure in their growing points may be the very signs of sentience. Perhaps the toyons and bay laurels are aware of the sun that their branches seek.

We see the homely umber of a winter wren flitting in the shadowed undergrowth. A Wilson's warbler, black-capped and brilliant yellow, darts and picks web worms in the leaves above. On a fallen bay, a rufous-sided towhee trills "drink your tea," as its mate turns over leaves in the duff. These delicate creatures are the descendants of the mighty dinosaurs. Are these lifeforms conscious? How could it be otherwise? Like the dinosaurs, the birds have brains similar to ours in basic pattern. Nerve impulses travel along the same ancient paths in all of us, and our brains are bathed in the same kinds of hormones. The beauty that we hear in their songs tells us that at some level our consciousnesses must work in similar ways.

As we go deeper into the woods, darkness gathers. Douglas firs and redwoods tower above us; the balance of life is high overhead. It is quiet; only a creeper spirals up a redwood trunk. The fir is panrooted and penetrates only a few feet into the subsoil; the redwood roots go only a little farther into the volcanic bedrock. We are walking at the very base of life. The energy from the litter of the sunlight is but slowly released: it takes years for the bacteria and fungi of the forest floor to decay the fallen needles of the tall conifers.

It is quiet as we pass, and our muffled footfalls do not wake the deermice dreaming in their dens. The waves of potential that sweep their brains are like ours; the hormones that manage their sleep, their sexuality, their attachments and animosities are virtually the same as our own; their runways in the litter are like the paths we are following today. Anyone who has patiently watched mice knows how like us they can be. Mice are small mirrors for our mood and mind; in their bodies and movements we can recognize ourselves without a veil of words.

We move upwards and the woods lighten. We ascend into a natural parkland with low-boughed trees scattered among the grass. We stop to feel the smooth red bark of the madrone and the purple bark of the manzanita. We see the yellow clumps of mule ears and the paired pink flowers of the honeysuckles. Although colors do not exist in the external world, they paint the internal world of many lifeforms. The yellow of the mule ears is a message for the bee, and the pink of the honeysuckle has meaning for the hummingbird. It is true that we all must walk along the impassive surface of material reality; yet the bright colors we see can be signals that there is more beyond.

A red admiral feeds at the mule ears, its satin-black wings bisected by a red bar, bearing bright white spots on its leading wings like epaulets. What could be the consciousness that wears such a dashing naval uniform on its fluttering voyages between the bright flowers in the bleaching grasses? This butterfly shares with us the most basic element of our nervous system: the neuron. Yet the butterfly's neurons are joined in such a radically contrasting configuration to our own. Similarities are our evidence for consciousness, but what can we make of differences?

We climb a rusting wire fence and come out in the open. Almost all life lies below us, in the mix of subclovers, rye, and wild oats, or immediately below it in the thin soil. Above the pasture, the air holds only microscopic forms of life in suspension; these are dwarfed by a turkey vulture, whose gliding six-foot wingspan suggests a cruising pterodactyl. A flock of sheep studies us and begins to move aside. We are walking through the fearful reality of the sheep. Their brains are like our own, their hearts are beating faster, their adrenaline and noradrenaline are flowing. We pass through a universe of feeling.

Invisible and visible, we walk on and the sheep again begin to graze, the vulture wheels, and the grass bends beneath our feet.

The invisible world of consciousness is the ghost of these sunny kingdoms of plants and animals. The sun is recreated in sugars and starches, in bone and sinew, in thought and mind. We are all the reflected glory of the sun.

Look at the horizon. To the west is the Pacific; to the east, the snow caps of the distant Sierra. The seed heads of the wild oats dip in the light wind. A swallow weaves in the air above the sheep. We are on top of Sonoma Mountain.

BIBLIOGRAPHY

ALCOCK, JOHN. *Animal Behavior: An Evolutionary Approach.* Sunderland, Mass.: Sinauer, 1975.

ALLISON, TRUETT J., and DOMINIC V. CICONETTI. "Sleep in Mammals: Ecological and Constitutional Correlates." *Science* 194 (1976): 732–34.

ALLISON, TRUETT J., and HENRY VAN TWYVER. "The Evolution of Sleep." *Natural History* 79 (February 1970): 56–65.

ARISTOTLE. *Historia Animalium,* Vol. 4, Book 9, 620b. Translated by D'Arcy Wentworth Thompson. Oxford: Clarendon, 1910.

BAKER, C. SCOTT, and LOUIS M. HERMAN. "Where Whales Go to Extremes." *Natural History* 94 (October 1985): 52–61.

BENTHRUP, F. W. "Reception and Transduction of Electrical and Mechanical Stimuli." *Encyclopedia of Plant Physiol-*

ogy. Vol. 7, *Physiology of Movements.* Edited by W. Haupt and M. E. Feinleib. Berlin: Springer-Verlag, 1979.

BLAKEMORE, RICHARD P., and RICHARD B. FRANKEL. "Magnetic Navigation in Bacteria." *Scientific American* 245 (December 1981): 58–65.

BRAZIER, MARY A. B. *Electrical Activity of the Nervous System.* 4th ed. Baltimore: Williams and Wilkens, 1978.

BRELAND, KELLER. *Animal Behavior.* New York: Macmillan, 1966.

———, and MARIAN BRELAND. "The Misbehavior of Organisms." *American Psychologist* 16 (1961): 681–84.

BROWN, JASON. *Mind, Brain, and Consciousness: Neurophysiology of Cognition.* New York: Academic Press, 1977.

COLLETT, T. S. "Sensory Guidance of Motor Behavior." In *Animal Behavior.* Vol. 1, *Causes and Effects,* edited by T. R. Halliday and P. J. Slater. New York: W. H. Freeman, 1983.

COLLIAS, NICHOLAS, and ELSIE C. COLLIAS, eds. *External Construction by Animals.* Stroudsburg, Pa: Dowden, Hutchinson, and Ross, 1976.

CONFUCIUS. *The Analects, the Great Learning, and the Doctrine of the Mean.* Translated by J. Legge. New York: Dover, 1971.

CORNING, WILLIAM C., JAMES A. DYAL, and A. O. D. WILLOWS, eds. *Invertebrate Learning.* Vols. 1, 2, and 3. New York: Plenum, 1973–75.

COUSTEAU, JACQUES-YVES, and PHILIPPE DIOLÉ. *Octopuses and Squid: The Soft Intelligence.* New York: Doubleday, 1973.

DARLING, JAMES D., KIMBERLEY M. GIBSON, and GREGORY K. SILBER. "Observations of the Abundance and Behavior of Humpback Whales (*Megaptera novaeangliae*) off West Maui 1977–79." In *Communication and Behavior of Whales,* edited by Roger N. Payne. AAAS Selective Symposium 76. Boulder: Westview Press, 1983.

DARWIN, CHARLES. *The Expression of the Emotions in Man and Animals.* London: Murray, 1872.

———. *The Power of Movement in Plants.* New York: Appleton, 1897.

DAVIS, JOEL. *Endorphins: New Waves in Brain Chemistry.* New York: Dial Press, 1984.

DEMENT, WILLIAM C. *Some Must Watch While Some Must Sleep.* New York: W. H. Freeman, 1978.

DESCARTES, RENÉ. "Letter to the Marquis of Newcastle." In *Descartes Selections,* edited by Ralph M. Eaton. New York: Scribner's, 1927.

DETHIER, VINCENT G. "Whose Real World?" *American Zoologist* 9 (1969): 241–49.

DURDEN-SMITH, JO, and DIANE DESMOND. *Sex and the Brain.* New York: Arbor House, 1983.

EMLEN, STEPHEN T. "Bird Migration: Influence of Physiological State upon Celestial Orientation." *Science* 165 (1969): 716–18.

———. "The Stellar-Orientation System of a Migratory Bird." *Scientific American* 233 (August 1975): 102–11.

———, and JOHN T. EMLEN. "A Technique for Recording Migratory Orientations of Captive Birds." *Auk* 83 (1966): 361–67.

FREUD, SIGMUND. *The Interpretation of Dreams.* In Vol. 4 of *The Standard Edition of the Complete Psychological Works of Sigmund Freud.* Translated by James Strachey. London: Hogarth Press and the Institute of Psycho-analysis, 1953.

GALSTON, ARTHUR W. *Green Wisdom.* New York: Basic Books, 1981.

GARDNER, R. ALLAN, and BEATRICE T. GARDNER. "Teaching Sign Language to a Chimpanzee." *Science* 165 (1969): 664–72.

GILLIARD, E. THOMAS. "The Evolution of Bowerbirds." *Scientific American* 209 (August 1963): 38–46.

GOULD, JAMES L., and C. G. GOULD. "The Insect Mind: Physics or Metaphysics." In *Animal Mind—Human Mind,* edited by Donald R. Griffen. New York: Springer-Verlag, 1982.

GOULD, S. J. " The Ediacarian Experiment." *Natural History* 14 (February 1984): 14–23.

GRIFFIN, DONALD R. *The Question of Animal Awareness.* New York: Rockefeller University Press, 1976.

————. *The Question of Animal Awareness.* 2d ed., rev. and enl. New York: Rockefeller University Press, 1981.

————, ed. *Animal Mind—Human Mind.* New York: Springer-Verlag, 1982.

————. *Animal Thinking.* Cambridge, Mass.: Harvard University Press, 1984.

GUILLEMIN, ROGER, and ROGER BURGUS. "Hormones of the Hypothalamus." *Scientific American* 227 (November 1972): 24–33.

GUTHRIE, R. D. "A New Theory of Mammalian Rump Patch Evolution." *Behavior* 38 (1971): 132–45.

HAYES, CATHERINE. *The Ape in Our House.* New York: Harper & Row, 1951.

HILLER, BARBARA. "Conversations with Koko." *Gorilla* 8, no. 1 (1984b): 8–9.

————. "Dinosaur Trouble and the Last Laugh." *Gorilla* 8, no. 2 (1984a): 10.

HIPPOCRATES. "The Sacred Disease." In *Hippocrates,* Vol. 2, translated by W. H. Jones. Cambridge, Mass.: Harvard University Press, 1952.

HOBSON, J. ALLAN. "Electroencephalographic Correlates of Behavior in Tree Frogs." *Nature* 220 (1968): 386–87.

HOPKINS, ANTHONY. *Epilepsy: The Facts.* New York: Oxford University Press, 1981.

HOPKINS, CARL D. "Electric Communication in Fish." *American Scientist* 62 (1974): 426–37.

————. "On the Diversity of Electric Signals in a Community of Mormyrid Electric Fishes of West Africa." *American Zoologist* 21 (1981): 211–22.

JERISON, HARRY J. "Brain Evolution and Dinosaur Brains." *American Naturalist* 103 (1969): 575–88.

————. *Evolution of the Brain and Intelligence.* New York: Academic Press, 1973.

————. "Paleoneurology and the Evolution of Mind." *Scientific American* 234 (January 1976): 90–101.

————. "Evolution of the Brain." In *The Human Brain,* edited by M. C. Wittrock *et al.* Englewood Cliffs, N. J.: Prentice-Hall, 1977.

JOLLY, ALISON. "A New Science That Sees Animals as Conscious Beings." *Smithsonian* 15 (March 1985): 67–75.

JOSEPHSON, ROBERT K. "The Response of a Hydroid to Weak Water Disturbances." *Journal of Experimental Biology* 38 (1961): 17–27.

JOUVET, MICHEL. "States of Sleep." *Scientific American* 216 (February 1967): 62–72.

————. "The Function of Dreaming: A Neurophysiologist's Point of View." In *Handbook of Psychobiology,* edited by Michael S. Gazzaniga and Colin Blakemore. New York: Academic Press, 1975.

KAFKA, FRANZ. *Parables and Paradoxes.* New York: Schocken Books, 1958.

KAVALIERS, MARTIN, MAURICE M. HIRST, and G. CAMPBELL TES-KEY. "A Functional Role for an Opiate System in Snail Thermal Behavior." *Science* 220 (1983): 99–100.

KOLATA, GINA. "Birds, Brains, and the Biology of Song." *Science 85* (December 1985): 58–63.

KREBS, JOHN R., RUTH ASHCROFT, and MIKE I. WEBBER. "Song Repertoires and Territory Defence in the Great Tit." *Nature* 271 (1978): 539–42.

KROODSMA, DONALD E. "Reproductive Development in a Female Songbird." *Science* 192 (1976): 574–75.

LAO TSU. *Tao Te Ching.* Translated by Gia-fu Feng and Jane English. New York: Vintage, 1972.

McGREGOR, PETER K., JOHN R. KREBS, and CHRISTOPHER M. PERRINS. "Song Repertoires and Lifetime Reproductive Success in the Great Tit *(Parus major)*." *American Naturalist* 118 (1981): 149–59.

MacLEAN, PAUL D. "The Brain in Relation to Empathy and Medical Education." *Journal of Nervous and Medical Education* 144 (1967): 372–82.

————. "On the Origin and Progressive Evolution of the Triune Brain." In *Primate Brain Evolution,* edited by Este Armstrong and Dean Falk. New York: Plenum, 1982.

MARSHALL, A. J. *Bowerbirds, Their Displays and Breeding Cycles.* Oxford: Clarendon, 1954.

MONTAIGNE, MICHEL DE. *The Essays of Montaigne.* Translated by J. Florio. New York: Ames, 1967.

MORGAN, C. LLOYD. *An Introduction to Comparative Psychology.* London: Walter Scott, 1894. (Facsimile in *Significant Contributions to the History of Psychology,* Series D, Vol. 2. Washington, D. C.: University Publications of America, 1977.)

————. *Habit and Instinct.* London: Arnold, 1896.

NAITOH, YUTAKA, and ROGER ECKERT. "Ionic Mechanisms Controlling Behavioral Responses of Paramecium to Mechanical Stimulation." *Science* 164 (May 1969): 963–65.

NEWMAN, ERIC A., and PETER H. HARTLINE. "Infrared Vision of Snakes." *Scientific American* 246 (March 1982): 116–127.

NOTTEBAUM, FERNANDO, and ARTHUR P. ARNOLD. "Sexual Discrimination in Vocal Control Areas of the Songbird Brain." *Science* 194 (1976): 211–13.

PATTERSON, FRANCINE G., and EUGENE LINDEN. *The Education of Koko.* New York: Holt, Rinehart and Winston, 1981.

PATTERSON, FRANCINE G., and TONI PETERSON. "Free Word Association by a Gorilla." *Gorilla* 7, no. 2 (1984): 7.

PAVLOV, IVAN P. *Conditioned Reflexes.* Edited by G. A. Anrep. New York: Dover, 1960.

PAYNE, KATHERINE, PETER TYACK, and ROGER N. PAYNE. "Progressive Changes in the Songs of Humpback Whales (*Megaptera novaeangliae*): A Detailed Analysis of Two Seasons in Hawaii." In *Communication and Behavior of Whales,* edited by Roger N. Payne. AAAS Selected Symposium 76. Boulder: Westview Press, 1983.

PAYNE, ROGER N., ed. *Communication and Behavior of Whales.* AAAS Selected Symposium 76. Boulder: Westview Press, 1983.

————, and LINDA L. GUINEE. "Humpback Whale (*Megaptera novaeangliae*) Songs as an Indicator of 'Stocks.'" In *Communication and Behavior of Whales,* edited by Roger N.

Payne. AAAS Selected Symposium 76. Boulder: Westview Press, 1983.

PENFIELD, WILDER. *The Mystery of the Mind.* Princeton: Princeton University Press, 1975.

———, and T. RASMUSSEN. *The Cerebral Cortex of Man.* New York: Macmillan, 1950.

PEPPERBERG, I. M. "Functional Vocalization by an African Gray Parrot *(Psittacus erithacus)*" *Zeitschrift fuer Tierpsychologie* 55 (1981): 139–60.

———. "Cognition in the African Gray Parrot: Preliminary Evidence for (Auditory / Vocal) Comprehension of the Class Concept." *Animal Learning and Behavior* 11 (1983): 179–185.

PRIMBRAM, KARL H. *Languages of the Brain: Experimental Paradoxes and Principles of Neuropsychology.* Englewood Cliffs, N.J.: Prentice-Hall: 1971.

RACUSEN, RICHARD H., and RUTH L. SATTER. "Rhythmic and Phytochrome-Regulated Changes in Transmembrane Potential in *Samanea Pulvini.*" *Nature* 255 (1975): 408–10.

ROEDER, KENNETH D. *Nerve Cells and Insect Behavior.* Cambridge, Mass.: Harvard University Press, 1967.

ROMANES, GEORGE J. *Mental Evolution in Animals.* New York: Ames Press, 1969. (Original ed.—London: Vegan Paul, Trench & Co., 1884.)

———. *Mental Evolution in Man.* New York: Appleton, 1889.

SANDERS, G. D. "The Cephalopods." In *Invertebrate Learning,* edited by William C. Corning, James A. Dyal, and A.O.D. Willows. New York: Plenum, 1975.

SATTER, RUTH L. "Leaf Movements and Tendril Curling." In *Encyclopedia of Plant Physiology,* Vol. 7, *Physiology of Movements,* edited by W. Haupt and M. E. Feinleib. Berlin: Springer-Verlag, 1979.

SCHULTZ, JACK C. "Tree Tactics." *Natural History* 92 (May 1983): 12–25.

SCOTT, BRUCE I. "Electricity in Plants." *Scientific American* 207 (October 1962): 107–16.

SEARCY, WILLIAM A., and PETER MARLER. "A Test for Responsiveness to Song Structure and Programming in Female Sparrows." *Science* 213 (1981): 926–28.

SHETTLEWORTH, SARA J. "Memory in Food-Hoarding Birds." *Scientific American* 248 (March 1983): 102–110.

SKINNER, B. F. *Behavior of Organisms.* New York: Appleton, 1938.

——. *Science and Human Behavior.* New York: Crowell-Collier-Macmillan, 1953.

SLACK, ADRIAN. *Carnivorous Plants.* Cambridge, Mass.: MIT Press, 1980.

SMITH, SUSAN J. "The Tiniest Established Permanent Floating Crap Game in the Northeast." *Natural History* 94 (March 1985): 42–47.

SMITH, W. JOHN. "Message, Meaning, and Context in Ethology." *American Naturalist* 99 (1965): 405–07.

——. *The Behavior of Communicating.* Cambridge, Mass.: Harvard University Press, 1977.

SUTPHEN, JOHN. "Body State Communication Among Cetaceans." In *Mind in the Water,* edited by Joan McIntyre. New York: Scribner's, 1974.

THOMAS. *Gospel According to Thomas.* Translated by A. Guillaumont, H-Ch. Puech, G. Quispel, W. Till, and Yassah Àbd al Masih. New York: Harper & Row, 1959.

TINBERGEN, NIKO. "Ueber die Orientierung des Bienenwolfes (*Philanthus triangulum*) Fabr" ["On the Orientation of the Digger Wasp (*Philanthus triangulum*) Fabr"]. *Zeitschrift fuer Vergleichende Physiologie* 16 (1932): 305–54. In *The Animal in Its World,* Vol. 1, edited by Niko Tinbergen. Cambridge, Mass.: Harvard, 1972.

TOLMAN, EDWARD C. *Purposive Behavior in Animals and Men.* New York: Century, 1932.

TOMPKINS, PETER, and CHRISTOPHER BIRD. *The Secret Life of Plants.* New York: Harper & Row, 1973.

TYACK, PETER. "Interactions Between Singing Hawaiian Humpback Whales and Conspecifics Nearby." *Behavioral Ecology and Sociobiology* 8 (1981): 105–16.

VAN HOOFF, J. A. R. A. M. "The Facial Displays of the Catarrhine

Monkeys and Apes." In *Primate Ethology,* edited by Desmond Morris. Chicago: Aldine, 1967.

———. "A Comparative Approach to the Phylogeny of Laughter". In *Nonverbal Communication,* edited by Robert A. Hinde. Cambridge: Cambridge University Press, 1972.

VAN TWYVER, HENRY, and TRUETT H. ALLISON. "A Polygraphic Behavioral Study in the Pigeon (*Columba livia*)." *Experimental Neurology* 35 (1972): 138–52.

WALKER, STEPHEN F. *Animal Thought.* London: Routledge & Kegan Paul, 1983.

WARSHALL, PETER. "The Ways of Whales." In *Mind in the Waters,* edited by Joan McIntyre. New York: Scribner's, 1974.

WASHBURN, MARGARET F. *Animal Mind.* 3d ed. New York: Macmillan, 1926.

WATERHOUSE, DENNIS, ANNA MINDESS, and ROBIN SILBERING. "Understanding Sign Language." *Gorilla* 8, no. 1 (1984): 7–9.

WATSON, JOHN B. *Behavior: An Introduction to Comparative Psychology.* New York: Henry Holt and Co., 1914.

———. *Psychology from the Standpoint of a Behaviorist.* Philadelphia: Lippincott, 1919.

———. *Behaviorism.* Rev. ed. New York: Norton, 1930.

WELLS, MARTIN J. *Octopus.* London: Chapman and Hall, 1978.

WILDER, M. BYRON, GLENN R. FARLEY, and ARNOLD STARR. "Endogenous Late Positive Component of the Evoked Potential in Cats Corresponding to P300 in Humans." *Science* 211 (1981): 605–7.

WOLFE, THOMAS. *Look Homeward Angel.* New York: Scribner's, 1936.

WU, CHAU H. "Electric Fish and the Discovery of Animal Electricity." *American Scientist* 72 (1984): 598–607.

YOUNG, JOHN Z. *A Model of the Brain.* Oxford: Oxford University Press, 1964.

INDEX

Animal kingdom *(cont'd)*
  nerve cords and, 37, 39
  nerve nets and, 37
  nervous system and, *see*
    Nervous system
  octopuses and, 41–44
  praying mantises and, 39–41
  single-celled, 35–36
  voltage pulses in, 23, 28, 30
Animal Mind (Washburn), 5,
  12–13
Animal Mind—Human Mind
  (Griffin), 11
Animal Thinking (Griffin), 11
Animal Thought (Walker), 11
Aristotle, 115, 125
Ashcroft, Ruter, 138
Autonomic nervous system, 86,
  88
Auxin, 19–21, 24
Awareness, 14
Axon, 26

Baboons, 114, 143–44
Backster, Cleve, 21–22
Bacteria, 127–28
Baker, C. Scott, 134
Barbary apes, 143–44
Bared-teeth expression, 143–45
  relaxed open-mouth, 144
  silent, display, 143–44
  vocalized, threat, 143
Barn swallows, 9–10, 34
Basal nucleus, 89, 90
Bats, 11, 76, 122
Bees, 113, 121
Beewolf, 112
Behavior, 30
Behavior as an expression of
    consciousness, 102–18
  elementary tests of, 104–105
  evidence for, 116–18
  exploration, 105, 107–108

fraction of consciousness
    and, 102–103
  inferring mind and, 103
  language, 113–15
  learning, *see* Learning
  levels of consciousness, 105
  man and animals, 115–16
  orientation, 105, 107
  scala naturae and, 115–16
  sensitivity, 105, 106
  tree of life and, 116
Behaviorism, 10, 12, 70,
  111–12, 140
Benthrup, F. W., 23
Berger, Hans, 72–73
Bible, 152
Binary code, 4
Biology, 10–13
  revolution in, 11–13
Biopotential, 22–25
Bird, Christopher, 21
Birds, 113, 115
  brain sizes of, 60, 61
  communication with, 145–46
  consciousness of, 153
  culture and, 114
  migration of, 131–33
  prehistoric era, 61, 62
  sleep of, 75, 76
  songs of, 135–39
"Black substance," 90
Bladderworts, 18
Blakemore, Richard P., 127
Blood pressure, 94
"Blue place," 88
Bottle-nosed dolphin, 63
Brain:
  consciousness and, 46–47,
    51–60, 70–71
  described, 47–51
  evolution of, *see* Evolution
    of the brain
  hormones and, *see*
    Hormones